Handbook of Ceramic Materials

Handbook of Ceramic Materials

Iker Morris

NY RESEARCH PRESS
New York

Published by NY Research Press
118-35 Queens Blvd., Suite 400,
Forest Hills, NY 11375, USA
www.nyresearchpress.com

Handbook of Ceramic Materials
Iker Morris

International Standard Book Number: 978-1-63238-867-4 (Hardback)

Cataloging-in-Publication Data

Handbook of ceramic materials / Iker Morris.
p. cm.
Includes bibliographical references and index.
ISBN 978-1-63238-867-4
1. Ceramic materials. 2. Mines and mineral resources. I. Morris, Iker.
TA455.C43 H36 2022
620.14--dc23

Contents

Preface

A ceramic is an inorganic non-metallic compound which exists in solid state. It could be made up of metal, metalloid or non-metal atoms which are bonded through ionic and covalent bonds. Ceramic materials are usually hard and brittle. They are not easily corroded by acidic or caustic environments and can generally withstand large amounts of heat. Ceramics are classified into two types on the basis of materials-crystalline ceramics and non-crystalline ceramics. Crystalline ceramics are difficult to process. Non-crystalline ceramics are easier to process since they are formed from melts. Thus, they can be melted and then cast into desired shapes. A few applications of ceramics are carbon-ceramic brake disks for vehicles, ceramic ball bearings and bio-ceramics. The objective of this book is to give a general view of the different areas of ceramics. Some of the diverse topics covered in it address the varied branches that fall under this category. This book will prove to be immensely beneficial to students and researchers in this field.

A short introduction to every chapter is written below to provide an overview of the content of the book:

Chapter 1 - A solid material which is composed of an inorganic compound of metal, non-metal or metalloid atoms is known as a ceramic. The atoms within a ceramic are mostly held in ionic and covalent bonds. This chapter will provide a brief introduction to all the significant aspects of ceramics such as their properties and structure; **Chapter 2 -** Some of the diverse types of ceramics are traditional ceramics, glass-ceramics, ceramic cements, structural ceramics, advanced ceramics, alumina ceramics, electroceramics and piezoelectric ceramics. This chapter has been carefully written to provide an easy understanding of these types of ceramics; **Chapter 3** - The branch of engineering which is involved in creating objects from non-metallic and inorganic materials is known as ceramic engineering. Some of the applications of ceramics are ceramic inks, ceramic tiles and ceramic capacitors. The diverse aspects of ceramic engineering along with these applications have been thoroughly discussed in this chapter; **Chapter 4 -** There are numerous processes which are used in the formation of ceramics such as firing, extrusion, slip casting, dry pressing, ceramic mold casting, ceramic injection molding, tape casting and freeze casting. The topics elaborated in this chapter will help in gaining a better perspective about these ceramics forming processes; **Chapter 5 -** Ceramics are used in numerous manufacturing processes such as coating, machining and joining. Ceramic materials are also used to make cutting tools. The diverse applications of ceramics in these manufacturing processes as well as the hardness testing of ceramics have been thoroughly discussed in this chapter.

Finally, I would like to thank my fellow scholars who gave constructive feedback and my family members who supported me at every step.

Iker Morris

1

Introduction to Ceramics

A solid material which is composed of an inorganic compound of metal, non-metal or metalloid atoms is known as a ceramic. The atoms within a ceramic are mostly held in ionic and covalent bonds. This chapter will provide a brief introduction to all the significant aspects of ceramics such as their properties and structure.

Ceramics can be defined as heat-resistant, nonmetallic, inorganic solids that are (generally) made up of compounds formed from metallic and nonmetallic elements. Although different types of ceramics can have very different properties, in general ceramics are corrosion-resistant and hard, but brittle. Most ceramics are also good insulators and can withstand high temperatures. These properties have led to their use in virtually every aspect of modern life.

The two main categories of ceramics are traditional and advanced. Traditional ceramics include objects made of clay and cements that have been hardened by heating at high temperatures. Traditional ceramics are used in dishes, crockery, flowerpots, and roof and wall tiles. Advanced ceramics include carbides, such as silicon carbide, SiC; oxides, such as aluminum oxide, Al_2O_3; nitrides, such as silicon nitride, Si_3N_4; and many other materials, including the mixed oxide ceramics that can act as superconductors. Advanced ceramics require modern processing techniques, and the development of these techniques has led to advances in medicine and engineering.

Glass is sometimes considered a type of ceramic. However, glasses and ceramics differ in that ceramics have a crystalline structure while glasses contain impurities that prevent crystallization. The structure of glasses is amorphous, like that of liquids. Ceramics tend to have high, well-defined melting points, while glasses tend to soften over a range of temperatures before becoming liquids. In addition, most ceramics are opaque to visible light, and glasses tend to be translucent. Glass ceramics have a structure that consists of many tiny crystalline regions within a non-crystalline matrix. This structure gives them some properties of ceramics and some of glasses. In general, glass ceramics expand less when heated than most glasses, making them useful in windows, for wood stoves, or as radiant glass-ceramic cooktop surfaces.

Composition

Some ceramics are composed of only two elements. For example, alumina is aluminum oxide, Al_2O_3; zirconia is zirconium oxide, ZrO_2; and quartz is silicon dioxide, SiO_2. Other ceramic materials, including many minerals, have complex and even variable compositions. For example, the ceramic mineral feldspar, one of the components of granite, has the formula $KAlSi_3O_8$.

The chemical bonds in ceramics can be covalent, ionic, or polar covalent, depending on the chemical composition of the ceramic. When the components of the ceramic are a metal and a nonmetal, the bonding is primarily ionic; examples are magnesium oxide (magnesia), MgO, and barium titanate, $BaTiO_3$. In ceramics composed of a metalloid and a nonmetal, bonding is primarily covalent;

examples are boron nitride, BN, and silicon carbide, SiC. Most ceramics have a highly crystalline structure, in which a three-dimensional unit, called a unit cell, is repeated throughout the material. For example, magnesium oxide crystallizes in the rock salt structure. In this structure, Mg^{2+} ions alternate with O^{2-} ions along each perpendicular axis.

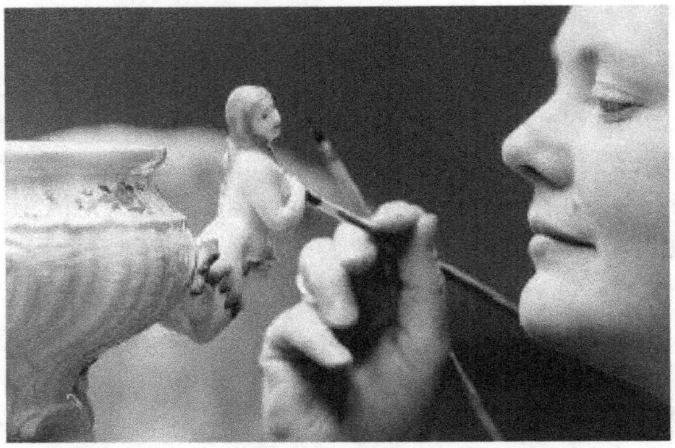

Ceramics are good insulators and can withstand high temperatures.
A popular use of ceramics is in artwork.

Manufacture of Traditional Ceramics

Traditional ceramics are made from natural materials such as clay that have been hardened by heating at high temperatures (driving out water and allowing strong chemical bonds to form between the flakes of clay). When artists make ceramic works of art, they first mold clay, often mixed with other raw materials, into the desired shape. Special ovens called kilns are used to "fire" (heat) the shaped object until it hardens.

Clay consists of a large number of very tiny flat plates, stacked together but separated by thin layers of water. The water allows the plates to cling together, but also acts as a lubricant, allowing the plates to slide past one another. As a result, clay is easily molded into shapes. High temperatures drive out water and allow bonds to form between plates, holding them in place and promoting the formation of a hard solid. Binders such as bone ash are sometimes added to the clay to promote strong bond formation, which makes the ceramic resistant to breakage. The common clay used to make flowerpots and roof tiles is usually red-orange because of the presence of iron oxides. White ceramics are made from rarer (and thus more expensive) white clays, primarily kaolin.

The oldest known ceramics made by humans are figurines found in the former Czechoslovakia that are thought to date from around 27,000 B.C.E. It was determined that the figurines were made by mixing clay with bone, animal fat, earth, and bone ash (the ash that results when animal bones are heated to a high temperature), molding the mixture into a desired shape, and heating it in a domed pit. The manufacture of functional objects such as pots, dishes, and storage vessels, was developed in ancient Greece and Egypt during the period 9000 to 6000 B.C.E.

An important advance was the development of white porcelain. Porcelain is a hard, tough ceramic that is less brittle than the ceramics that preceded it. Its strength allows it to be fashioned into beautiful vessels with walls so thin they can even be translucent. It is made from kaolin mixed with china stone, and the mixture is heated to a very high temperature (1,300 °C, or 2,372 °F). Porcelain was developed in China around C.E. 600 during the T'ang dynasty and was perfected during the

Ming dynasty, famous for its blue and white porcelain. The porcelain process was introduced to the Arab world in the ninth century; later Arabs brought porcelain to Spain, from where the process spread throughout Europe.

Bone china has a composition similar to that of porcelain, but at least 50 percent of the material is finely powdered bone ash. Like porcelain, bone china is strong and can be formed into dishes with very thin, translucent walls. Stoneware is a dense, hard, gray or tan ceramic that is less expensive than bone china and porcelain, but it is not as strong. As a result, stoneware dishes are usually thicker and heavier than bone china or porcelain dishes.

Manufacture of Advanced Ceramics

The preparation of an advanced ceramic material usually begins with a finely divided powder that is mixed with an organic binder to help the powder consolidate, so that it can be molded into the desired shape. Before it is fired, the ceramic body is called "green." The green body is first heated at a low temperature in order to decompose or oxidize the binder. It is then heated to a high temperature until it is "sintered," or hardened, into a dense, strong ceramic. At this time, individual particles of the original powder fuse together as chemical bonds form between them. During sintering the ceramic may shrink by as much as 10 to 40 percent. Because shrinkage is not uniform, additional machining of the ceramic may be required in order to obtain a precise shape.

Sol-gel technology allows better mixing of the ceramic components at the molecular level, and hence yields more homogeneous ceramics, because the ions are mixed while in solution. In the sol-gel process, a solution of an organometallic compound is hydrolyzed to produce a "sol," a colloidal suspension of a solid in a liquid. Typically the solution is a metal alkoxide such as tetramethoxysilane in an alcohol solvent. The sol forms when the individual formula units polymerize (link together to form chains and networks). The sol can then be spread into a thin film, precipitated into tiny uniform spheres called microspheres, or further processed to form a gel inside a mold that will yield a final ceramic object in the desired shape. The many crosslinks between the formula units result in a ceramic that is less brittle than typical ceramics.

Although the sol-gel process is very expensive, it has many advantages, including low temperature requirements; the ceramist's ability to control porosity and to form films, spheres, and other structures that are difficult to form in molds; and the attainment of specialized ceramic compositions and high product purity.

Porous ceramics are made by the sol-gel process. These ceramics have sponge like structures, with many porelike lacunae, or openings, that can make up from 25 to 70 percent of the volume. The pore size can be large, or as small as 50 nanometers (2×10^{-6} inches) in diameter. Because of the large number of pores, porous ceramics have enormous surface areas (up to 500 square meters, or 5,382 square feet, per gram of ceramic), and so can make excellent catalysts. For example, zirconium oxide is a ceramic oxygen sensor that monitors the air-to-fuel ratio in the exhaust systems of automobiles.

Aerogels are solid foams prepared by removing the liquid from the gel during a sol-gel process at high temperatures and low pressures. Because aerogels are good insulators, have very low densities, and do not melt at high temperatures, they are attractive for use in spacecraft.

Ceramic Composition and Properties

Ceramic composition and properties are the atomic and molecular nature of ceramic materials and their resulting characteristics and performance in industrial applications. Industrial ceramics are commonly understood to be all industrially used materials that are inorganic, nonmetallic solids. Usually they are metal oxides (that is, compounds of metallic elements and oxygen), but many ceramics (especially advanced ceramics) are compounds of metallic elements and carbon, nitrogen, or sulfur. In atomic structure they are most often crystalline, although they also may contain a combination of glassy and crystalline phases. These structures and chemical ingredients, though various, result in universally recognized ceramic-like properties of enduring utility, including the following: mechanical strength in spite of brittleness; chemical durability against the deteriorating effects of oxygen, water, acids, bases, salts, and organic solvents; hardness, contributing to resistance against wear; thermal and electrical conductivity considerably lower than that of metals; and an ability to take a decorative finish.

In chemical composition, for instance, diamond and graphite, which are two different forms of carbon, are considered to be ceramics even though they are not composed of inorganic compounds. There also are exceptions to the stereotypical properties ascribed to ceramics. To return to the example of diamond, this material, though considered to be a ceramic, has a thermal conductivity higher than that of copper—a property the jeweler uses to differentiate between true diamond and simulants such as cubic zirconia (a single-crystal form of zirconium dioxide). Indeed, many ceramics are quite conductive electrically. For instance, a polycrystalline (many-grained) version of zirconia is used as an oxygen sensor in automobile engines owing to its ionic conductivity. Also, copper oxide-based ceramics have been shown to have superconducting properties. Even the well-known brittleness of ceramics has its exceptions. For example, certain composite ceramics that contain whiskers, fibres, or particulates that interfere with crack propagation display flaw tolerance and toughness rivaling that of metals.

Nevertheless, despite such exceptions, ceramics generally display the properties of hardness, refractoriness (high melting point), low conductivity, and brittleness. These properties are intimately related to certain types of chemical bonding and crystal structures found in the material. Chemical bonding and crystal structure are addressed in turn below.

Chemical Bonds

Underlying many of the properties found in ceramics are the strong primary bonds that hold the atoms together and form the ceramic material. These chemical bonds are of two types: they are either ionic in character, involving a transfer of bonding electrons from electropositive atoms (cations) to electronegative atoms (anions), or they are covalent in character, involving orbital sharing of electrons between the constituent atoms or ions. Covalent bonds are highly directional in nature, often dictating the types of crystal structure possible. Ionic bonds, on the other hand, are entirely nondirectional. This nondirectional nature allows for hard-sphere packing arrangements of the ions into a variety of crystal structures, with two limitations. The first limitation involves the relative size of the anions and the cations. Anions are usually larger and close-packed, as in the face-centred cubic (fcc) or hexagonal close-packed (hcp) crystal structures found in metals.

These metallic crystal structures are illustrated in figure. Cations, on the other hand, are usually smaller, occupying interstices, or spaces, in the crystal lattice between the anions.

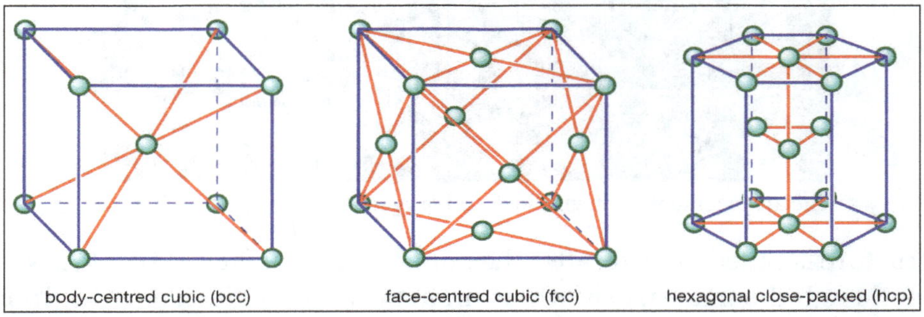

Three common metallic crystal structures.

The second limitation on the types of crystal structure that can be adopted by ionically bonded atoms is based on a law of physics—that the crystal must remain electrically neutral. This law of electroneutrality results in the formation of very specific stoichiometries—that is, specific ratios of cations to anions that maintain a net balance between positive and negative charge. In fact, anions are known to pack around cations, and cations around anions, in order to eliminate local charge imbalance. This phenomenon is referred to as coordination.

Most of the primary chemical bonds found in ceramic materials are actually a mixture of ionic and covalent types. The larger the electronegativity difference between anion and cation (that is, the greater the difference in potential to accept or donate electrons), the more nearly ionic is the bonding (that is, the more likely are electrons to be transferred, forming positively charged cations and negatively charged anions). Conversely, small differences in electronegativity lead to a sharing of electrons, as found in covalent bonds.

Secondary bonds also are important in certain ceramics. For example, in diamond, a single-crystal form of carbon, all bonds are primary, but in graphite, a polycrystalline form of carbon, there are primary bonds within sheets of crystal grains and secondary bonds between the sheets. The relatively weak secondary bonds allow the sheets to slide past one another, giving graphite the lubricity for which it is well known. It is the primary bonds in ceramics that make them among the strongest, hardest, and most refractory materials known.

Crystal Structure

Crystal structure is also responsible for many of the properties of ceramics. In figures through representative crystal structures are shown that illustrate many of the unique features of ceramic materials. Each collection of ions is shown in an overall box that describes the unit cell of that structure. By repeatedly translating the unit cell one box in any direction and by repeatedly depositing the pattern of ions within that cell at each new position, any size crystal can be built up. In the first structure the material shown is magnesia (MgO), though the structure itself is referred to as rock salt because common table salt (sodium chloride, NaCl) has the same structure. In the rock salt structure each ion is surrounded by six immediate neighbours of the opposite charge (e.g., the central Mg^{2+} cation, which is surrounded by O^{2-} anions). This extremely efficient packing allows for local neutralization of charge and makes for stable bonding. Oxides that crystallize in this structure tend to have relatively high melting points. Magnesia, for example, is a common constituent in refractory ceramics.

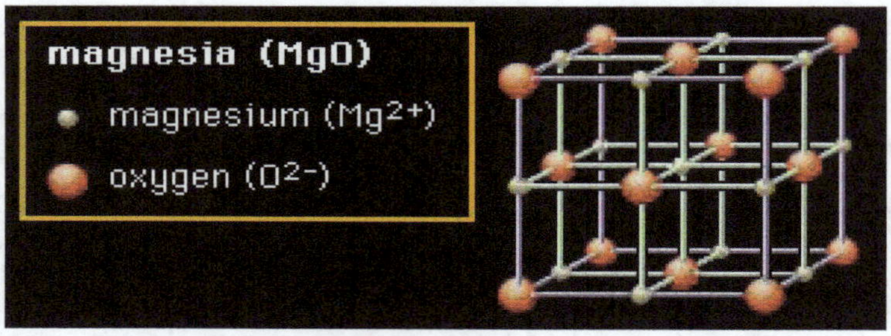

The arrangement of magnesium and oxygen ions in magnesia (MgO).

The second structure is called fluorite, after the mineral calcium fluoride (CaF_2), which possesses this structure—though the material shown, is urania (uranium dioxide, UO_2). In this structure the oxygen anions are bonded to only four cations. Oxides with this structure are well known for the ease with which oxygen vacancies can be formed. In zirconia (zirconium dioxide, ZrO_2), which also possesses this structure, a great number of vacancies can be formed by doping, or carefully inserting ions of a different element into the composition. These vacancies become mobile at high temperatures, imparting oxygen-ion conductivity to the material and making it useful in certain electrical applications. The fluorite structure also exhibits considerable open space, especially at the centre of the unit cell. In urania, which is used as a fuel element in nuclear reactors, this openness is believed to help accommodate fission products and reduce unwanted swelling.

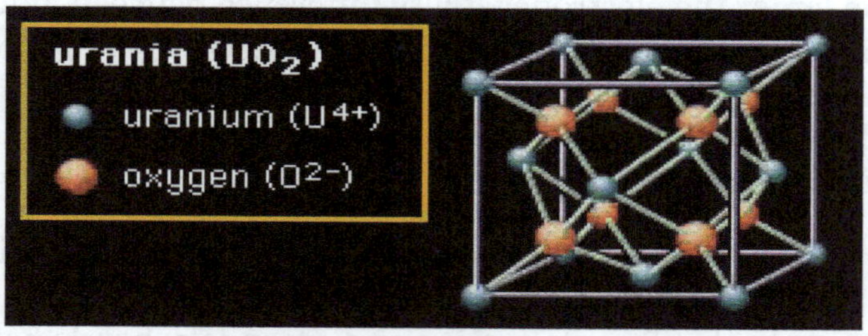

The arrangement of uranium and oxygen ions in urania (UO_2).

The third structure is called perovskite. In most cases the perovskite structure is cubic—that is, all sides of the unit cell are the same. However, in barium titanate ($BaTiO_3$), shown in the figure, the central Ti^{4+} cation can be induced to move off-centre, leading to a noncubic symmetry and to an electrostatic dipole, or alignment of positive and negative charges toward opposite ends of the structure. This dipole is responsible for the ferroelectric properties of barium titanate, in which domains of neighboring dipoles line up in the same direction. The enormous dielectric constants achievable with perovskite materials are the basis of many ceramic capacitor devices.

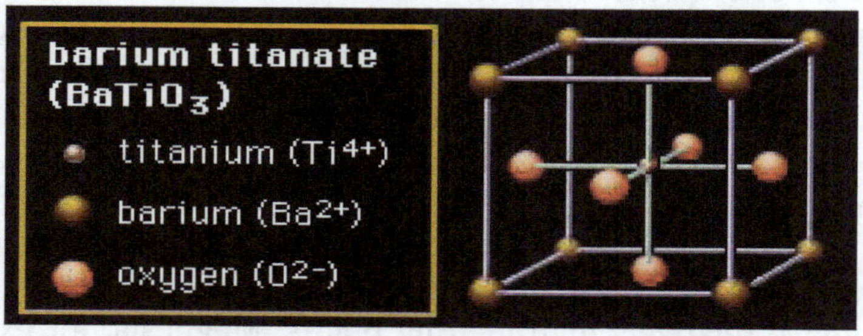

The arrangement of titanium, barium, and oxygen ions in barium titanate ($BaTiO_3$).

The non-cubic variations found in perovskite ceramics introduce the concept of anisotropy—i.e., an ionic arrangement that is not identical in all directions. In severely anisotropic materials there can be great variation of properties. These cases are illustrated by yttrium barium copper oxide (YBCO; chemical formula $YBa_2Cu_3O_7$), shown in figure. YBCO is a superconducting ceramic; that is, it loses all resistance to electric current at extremely low temperatures. Its structure consists of three cubes, with yttrium or barium at the centre, copper at the corners, and oxygen at the middle of each edge—with the exception of the middle cube, which has oxygen vacancies at the outer edges. The critical feature in this structure is the presence of two sheets of copper-oxygen ions, located above and below the oxygen vacancies, along which super conduction takes place. The transport of electrons perpendicular to these sheets is not favoured, making the YBCO structure severely anisotropic. One of the challenges in fabricating crystalline YBCO ceramics capable of passing large currents is to align all the grains in such a manner that their copper-oxygen sheets line up.

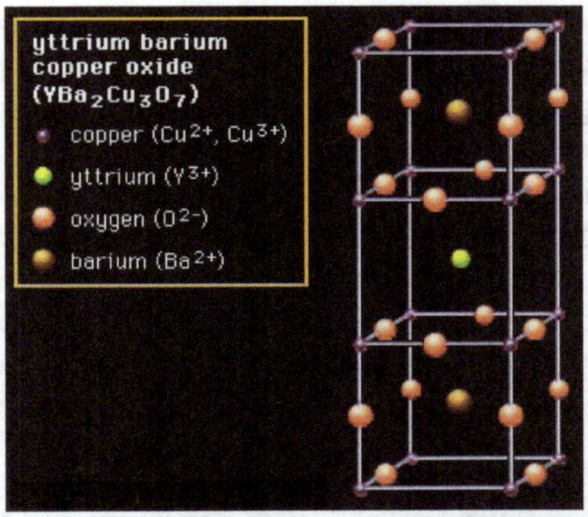

The arrangement of copper, yttrium, oxygen, and barium ions in yttrium bariumcopper oxide ($YBa_2Cu_3O_7$).

Nonconductivity

Ordinarily, ceramics are poor conductors of electricity and therefore make excellent insulators. Nonconductivity arises from the lack of "free" electrons such as those found in metals. In ionically bonded ceramics, bonding electrons are accepted by the electronegative elements, such as oxygen, and donated by the electropositive elements, usually a metal. The result is that all electrons are tightly bound to the ions in the structure, leaving no free electrons to conduct electricity. In covalent bonding, bonding electrons are similarly localized in the directional orbitals between the atoms, and there are no free electrons to conduct electricity.

There are two ways that ceramics can be made electrically conductive. At sufficiently high temperatures point defects such as oxygen vacancies can arise, leading to ionic conductivity. (This is pointed out in the case of zirconia.) In addition, the introduction of certain transition-metal elements (such as iron, copper, manganese, or cobalt), lanthanoid elements (such as cerium), or actinoid elements (such as uranium) can produce special electronic states in which mobile electrons or electron holes arise. The copper-based superconductors are a good example of conductive transition-metal oxide ceramics—in this case, conductivity arising at extremely low temperatures.

Brittleness

Unlike most metals, nearly all ceramics are brittle at room temperature; i.e., when subjected to tension, they fail suddenly, with little or no plastic deformation prior to fracture. Metals, on the other hand, are ductile (that is, they deform and bend when subjected to stress), and they possess this extremely useful property owing to imperfections called dislocations within their crystal lattices. There are many kinds of dislocations. In one kind, known as an edge dislocation, an extra plane of atoms can be generated in a crystal structure, straining to the breaking point the bonds that hold the atoms together. If stress were applied to this structure, it might shear along a plane where the bonds were weakest, and the dislocation might slip to the next atomic position, where the bonds would be re-established. This slipping to a new position is at the heart of plastic deformation. Metals are usually ductile because dislocations are common and are normally easy to move.

In ceramics, however, dislocations are not common (though they are not nonexistent), and they are difficult to move to a new position. The reasons for this lie in the nature of the bonds holding the crystal structure together. In ionically bonded ceramics some planes—such as the so-called (111) plane shown slicing diagonally through the rock salt structure in figure, top—contain only one kind of ion and are therefore unbalanced in their distribution of charges. Attempting to insert such a half plane into a ceramic would not favour a stable bond unless a half plane of the oppositely charged ion was also inserted. Even in the case of planes that were charge-balanced—for instance, the (100) plane created by a vertical slice down the middle of the rock salt crystal structure, as shown in figure, bottom—slip induced along the middle would bring identically charged ions into proximity. The identical charges would repel each other, and dislocation motion would be impeded. Instead, the material would tend to fracture in the manner commonly associated with brittleness.

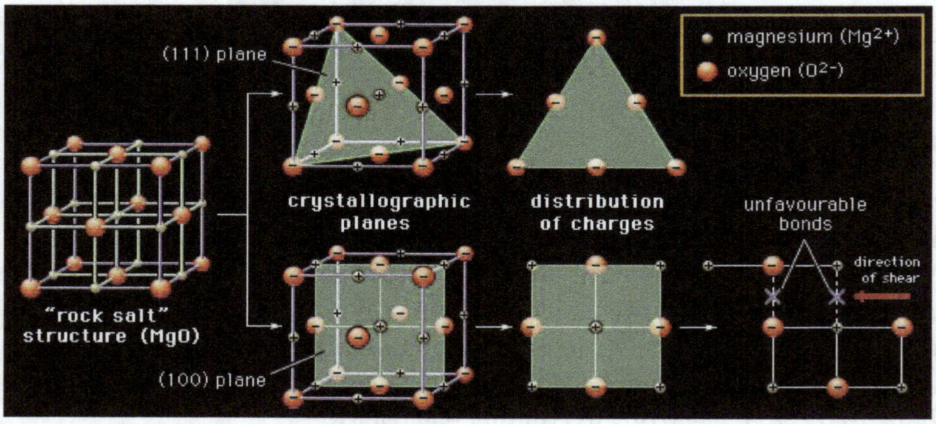

Figure shows barriers to slip in ceramic crystal structures. Beginning with the rock salt structure of magnesia (MgO; shown at left), in which there is a stable balance of positive and negative charges, two possible crystallographic planes show the difficulty of establishing stable imperfections. The (111) plane (shown at top) would contain atoms of identical charge; inserted as an imperfection into the crystal structure, such an imbalanced distribution of charges would not be able to establish a stable bond. The (100) plane (shown at bottom) would show a balance between positive and negative charges, but a shear stress applied along the middle of the plane would force identically charged atoms into proximity—again creating a condition unfavourable for stable bonding.

In order for polycrystalline materials to be ductile, they must possess more than a minimum number of independent slip systems—that is, planes or directions along which slip can occur. The presence of slip systems allows the transfer of crystal deformations from one grain to the next. Metals typically have the required number of slip systems, even at room temperature. Ceramics, however, do not, and as a result they are notoriously brittle.

Glasses, which lack a long-range periodic crystal structure altogether, are even more susceptible to brittle fracture than ceramics. Because of their similar physical attributes (including brittleness) and similar chemical constituents (e.g., oxides), inorganic glasses are considered to be ceramics in many countries of the world. Indeed, partial melting during the processing of many ceramics results in a significant glassy portion in the final makeup of many ceramic bodies (for instance, porcelains), and this portion is responsible for many desirable properties (e.g., liquid impermeability).

Powder Processing

Unlike metals and glasses, which can be cast from the melt and subsequently rolled, drawn, or pressed into shape, ceramics must be made from powders. As pointed out above, ceramics are seldom deformable, especially at room temperature, and the microstructural modifications achieved by cold-working and recrystallizing metals are impossible with most ceramics. Instead, ceramics are usually made from powders, which are consolidated and densified by sintering. Sintering is a process whereby particles bond and coalesce under the influence of heat, leading to shrinkage and reduction in porosity. A similar process in metal manufacturing is referred to as powder metallurgy.

Powder processing is used to make products that are normally identified as traditional ceramics—namely, white wares such as porcelain and china, structural clay products such as brick and tile, refractories for insulating and lining metallurgical furnaces and glass tanks, abrasives, and cements. It also is used in the production of advanced ceramics, including ceramics for electronic, magnetic, optical, nuclear, and biological applications. Traditional ceramics involve large volumes of product and relatively low value-added manufacturing. Advanced ceramics, on the other hand, tend to involve smaller volumes of product and higher value-added manufacturing.

Property of Ceramics

Physical Property

Physical properties are based on its crystal structure and chemical composition. Generally, Physical properties are identified by simple methods such as odor, color and physical form of the material (solid, liquid, gas).

Physical properties are based by its crystal structure, chemical composition and physical form of the ceramic-like solid, liquid and gas form also it depends upon the odor, color, and structure. In physical property, we consider the volume, density, heat capacity, melting and boiling point, the porosity of that material. These are a very important parameter of ceramic material.

Physical property includes that 3 main category:

- Density: Ceramics are intermediate in density between polymers (lower) and metals (higher) in the range of 2-6 gms/cm^3. Non-crystalline materials are less dense than crystalline ones. Compositions with several ALLOTROPES such as SiO2 will have minor differences in density.
- Melting and Boiling Point: Ceramic having high melting point i.e. 600-400 °C and also thermal conductivity is low and the value of thermal expansion are 1-15 ppm/C.
- Color: Depends on visible light interaction with "ions" or "pigments" (Color from ions [typically 0.2-0.4%] depends on the oxidation state).

Mechanical Property

Mechanical properties generally describe the strength of the material. It is important in structural and building materials. The mechanical property of mainly describe two main things:

- Hardness: Evaluating hardness homes depends heavily on the test technique. Knoop hardness is usually used for difficult, brittle substances. It uses a diamond-shaped tool to make a measurable impact on the cloth with a standardized load.
- Flexural Strength: Flexural strength is the effective power of a material under a bending load. The test takes under consideration grain shape and defects within the material and is typically performed on brittle substances. Evaluating flexural strength famous a cloth's sturdiness as it takes under consideration both compressive and tensile strengths.

Electrical Property

Electrical properties are related to how easily electrical energy passes through a material. Electricity effects of the ceramic material application that they used ceramic. The electrical properties of ceramic substrates perform an important task in the operation of electronic circuits.

The main considerations in electrical property are dielectric strength and resistivity

- Dielectric Strength: It is how a good deal electric ability, or voltage, a cloth can hold lower back before breaking down. Eventually, the voltage will become too much, and the material breaks right down to allow the electricity through.
- Electrical Resistivity: The electrical resistivity of a material is a measure of its ability to transport charge under the influence of an applied electric field.

Thermal Property

Thermal property means the ceramic material reacts to the change in temperature. In thermal properties of ceramic it mainly consists of two main parts i.e. thermal conductivity and coefficient of thermal expansion.

- Thermal Conductivity: It measures how well a fabric spreads warmness within itself. Cooking pans have high thermal conductivity permitting calmly distributed heat to pass speedily into the meals. On the opposite hand, insulative gloves are used to address warm objects because their low thermal conductivity prevents warmth from transmitting to sensitive arms.

- Coefficient of Thermal Expansion: On heating, ceramics expand this is generally known as thermal expansion. Thermal expansion is due to the conduction of heat. It defines how a whole lot a fabric expands based on outside temperatures. Most substances swell with the utility of warmth because the power reasons the atoms to move more hastily, stretching their bonds. Ceramics commonly have a low coefficient because of their sturdy interatomic bonds.

Magnetic Property

The magnetic property of any can be defined by how to respond to the material in the magnetic field. Ceramic are refractory material hence they can't respond for any magnetic field.

Chemical Property

Ceramics are bonded together by an ionic or covalent bond. Basically, these bonds result in good chemical resistance but have low thermal expansion, high melting point, and hardness. Ceramics are chemically inert, allowing them to be used in programs that cause different substances to degrade.

Dielectric Property

Dielectric properties are directly related to piezoelectric properties. The dielectric constant, which is the ratio of the permittivity of the material to its permittivity to free space, is an indicator of how the material concentrates on electric flux.

Because they can temporarily store electric energy dielectric ceramics serve an important role as a material for manufacturing capacitors, which, in turn, are responsible for properly supplying electric energy and signals to circuits.

Structure of Ceramic Materials

The following factors affect structure of ceramics:

- Balance of electrical charges of anions and cations.
- Radius Ratio (rc/ra).

Where,

- rc – radius of cation;
- ra – radius of anion.

Radius Ratio determines Coordination Number (CN) – the maximum number of anion nearest neighbors for a cation. The anion neighbors do not touch each other.

- $rc/ra = 0.225...0.414(SiO_2)$ CN = 4,
- $rc/ra = 0.414...0.732(SnO_2, PbO_2)$ CN = 6,
- $rc/ra = 0.732...1.0(ThO_2)$ CN = 8.

Covalent bonding component, which tends to form tetrahedral coordination, may affect the Coordination Number.

- Ions are packed with maximum density, providing minimum energy of the structure.

Table: Ceramic structures are classified and designated according to the pattern structures of several natural minerals:

Mineral Name	Formula	Coordination Number	Structure Characterization
Rocksalt	NaCl	6	Octahedral unit cell, cubic appearance
Zincblende	ZnS	4	FCC unit cell with S^{2-} anions at 4 tetrahedral sites
Fluorite	CaF_2	8-cation CN 4-anion CN	FCC unit cell with F^- anions at 8 tetrahedral sites
Corundum	Al_2O_3	6-cation CN 4-anion CN	HCP unit cell with O^{2-} anions at the lattice sites and Al^{3+} at interstitial sites
Perovskite	$CaTiO_3$	6-cation(Ti) CN 2-anion(O) CN	Cubic unit cell with Ti^{4+} cations coordinated octahedrally among six oxygen anions
Silicate	Combination of SiO_4^{4-} blocks	4	Tetrahedral arrangement with Si^{4+} cations at the center bonded to O^{2-} anions at the apices of the tetrahedron

Examples of Some Ceramic Structures

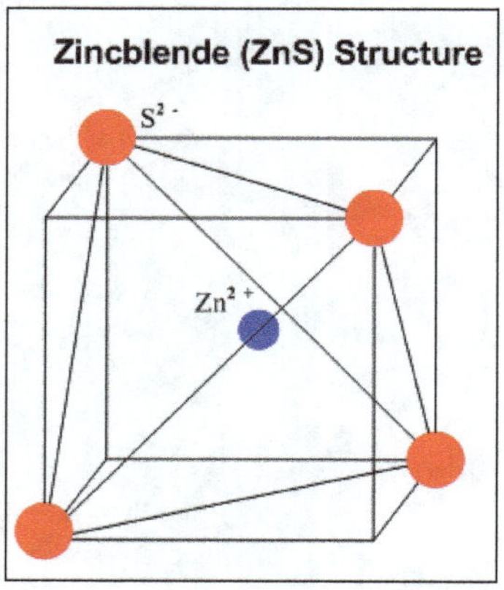

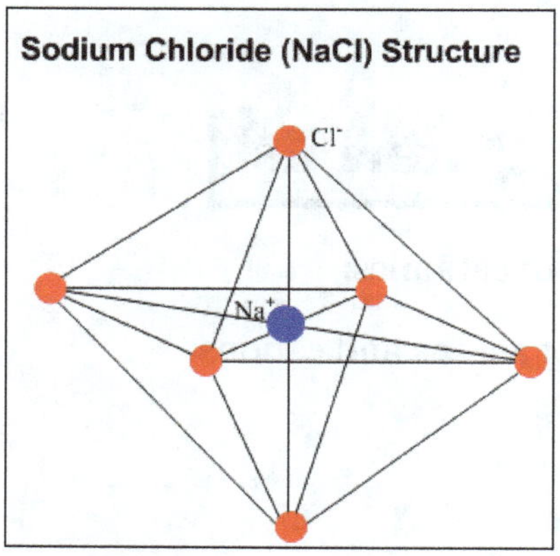

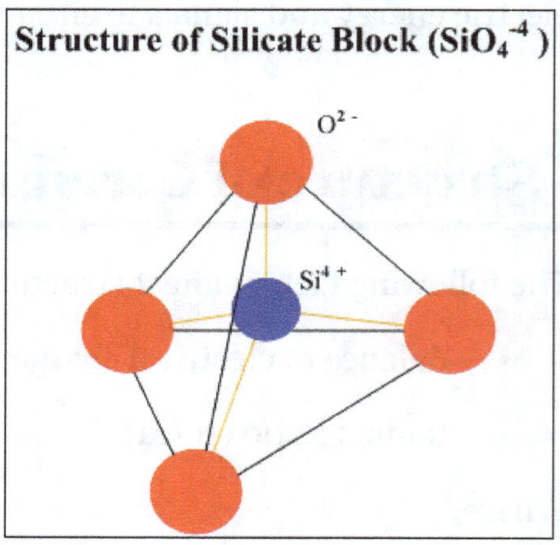

Tetrahedral silica block (SiO_4^{-4}) may form various silicate structures:

- Island and Double Island Silicates: Single or two silica blocks, containing other cations, form Island (olivine) or Double Island Silicates (hemimorphite).

- Ring and chain structures: Several silica units join each other, forming a chain (orthopyroxenes, clinopyroxenes, asbestos) or closed ring (beryl).
- Sheet (layer) structure: Silica units connect to each other, forming infinite sheet (micas, serpentine, chlorite, talc).
- Framework Silicate: Quartz, cristobalite, and tridymite minerals are based on the framework silicate structure.

Silicates exist in two forms: crystalline and amorphous (glasses).

Ionic and Covalent Bonding

Ceramics (ceramic materials) are non-metallic inorganic compounds formed from metallic (Al, Mg, Na, Ti, W) or semi-metallic (Si, B) and non-metallic (O, N, C) elements. Atoms of the elements are held together in a ceramic structure by one of the following bonding mechanism: Ionic Bonding, Covalent Bonding, Mixed Bonding (Ionic-Covalent).

Most of ceramic materials have a mixed bonding structure with various ratios between Ionic and Covalent components. This ratio is dependent on the difference in the electro-negativities of the elements and determines which of the bonding mechanisms is dominating ionic or covalent.

- Electronegativity,
- Ionic Bonding,
- Covalent Bonding,
- Ionic-Covalent (mixed) Bonding,
- Characterization of ceramics properties.

Electronegativity

Electronegativity is an ability of atoms of the element to attract electrons of atoms of another element. Electronegativity is measured in a relative dimensionless unit (Pauling scale) varying in a range between 0.7 (francium) to 3.98 (fluorine).

Non-metallic elements are strongly electronegative. Metallic elements are characterized by low electronegativity or high electro-positivity – ability of the element to lose electrons.

Ionic Bonding

Ionic bonding occurs between two elements with a large difference in their electro-negativities (metallic and non-metallic), which become ions (negative and positive) as a result of transfer of the valence electron from the element with low electronegativity to the element with high electronegativity.

The typical example of a material with Ionic Bonding is sodium chloride (NaCl). Electropositive sodium atom donates its valence electron to the electronegative chlorine atom, completing its outer electron level (eight electrons).

As a result of the electron transfer the sodium atom becomes a positively charged ion (cation) and the chlorine atom becomes a negatively charged ion (anion). The two ions attract to each other by Coulomb force, forming a compound (sodium chloride) with ionic bonding. Ionic bonding is non-directional.

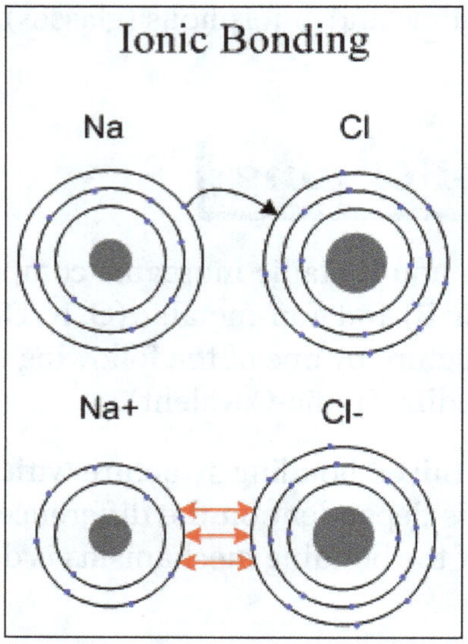

Covalent Bonding

Covalent bonding occurs between two elements with low difference in their electro-negativities (usually non-metallics), outer electrons of which are shared between the four neighboring atoms. Covalent Bonding is strongly directional. Ionic-Covalent (mixed) Bonding.

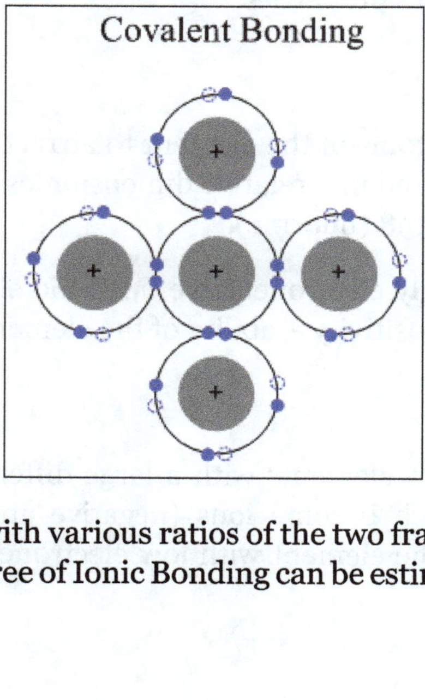

Ionic-covalent (mixed) bonding with various ratios of the two fractions (ionic and covalent) occurs in most of ceramic materials. Degree of Ionic Bonding can be estimated from the following formula:

$$I.F. = \exp(-0.25 * \Delta E^2)$$

Where,

- I.F. – fraction of ionic bonding;
- ΔE – difference in the electro-negativities of the elements.

Characterization of Ceramics Properties

In contrast to metallic bonding neither ionic nor covalent bonding form free electrons, therefore ceramic materials have very low electric conductivity and thermal conductivity. Since both ionic and covalent bonds are stronger than metallic bond, ceramic materials are stronger and harder than metals.

Strength of ionic and covalent bonds also determines high melting point, modulus of elasticity (rigidity), temperature and chemical stability of ceramic materials. Motion of dislocations through a ceramic structure is impeded therefore ceramics are generally brittle that limits their use as structural materials. Ceramics may have either crystalline or amorphous structure. There are also ceramic materials, consisting of two constituents: crystalline and amorphous.

Advantages and Disadvantages of Ceramics

Ceramic decor items, dishes and building materials add unique and personal touches to any home. The advantages of ceramic over other materials are numerous and although there are some downsides to the use of this material, as a whole it is a product which is extremely compatible with maintaining a clean, carefree and safe household and lifestyle.

The manufacturing of tile is one of the most well-known applications of ceramics.

Cost

Ceramics are usually relatively inexpensive, depending on the application. Installing ceramic tile throughout your home is far less expensive than adding carpet or hardwood flooring. There are cost differences though when ceramics are applied to the manufacturing of dishes and cookware. Some are extremely inexpensive while others demand top dollar because of design, brand and quality issues.

Color

Ceramic items also tend to hold color well. Unlike cement or stone, paint on ceramic vases, dishes and decorative items will stick to the surfaces and remain vivid for years to come. This makes ceramics also popular as an art and decorative form — interior decorators often opt to use ceramic both as floor and wall coverings, as well as in decorative pots, figurines, vases, lamps and knobs.

Durability

One of the downsides to ceramics is that the products are usually very fragile — not as fragile as their counterpart in porcelain but easily crushed, broken or shattered. Care should always be taken in the handling of ceramic items and cookware, and dishes should not be used if cracked or chipped. However, ceramics can handle extremes of heat and cold very well and are often used for freezer-to-oven casseroles and other favorite dishes.

Easy to Clean

Ceramics are commonly used in nursing homes, hospitals and child care facilities because of the high levels of sanitation with tile walls, flooring and ceramic fixtures. This is equally important in your own home, especially in areas such as the kitchen or bathroom. Ceramic is very easy to clean, stain resistant and does not retain the residues and smells of chemicals, foods or beverages.

Ceramic Surfaces

The texture of ceramic surfaces is cool and smooth to the touch and pleasing to the eye. It can be slippery, so care should be taken when handling ceramic dishes or cookware. Ceramic tile can be treated with an abrasive grit application available at your local building supply store or you can add texture to the tile surface to reduce accidents and ensure the safety of all who walk on it.

Water

Ceramics are especially useful in areas with heavy moisture such as bathrooms, kitchens and outdoors. Ceramic tile keeps these areas free from leakage, corrosion and mold. Ceramic dishes are leak-proof and are ideal for hot teas and other hot and cold beverages. Flower pots hold moist soil to aid in the growth of plants both indoors and out.

References

- Ceramics, Bo-Ce: chemistryexplained.com, Retrieved 31 March, 2019
- Ceramic-composition-and-properties: britannica.com, Retrieved 14 July, 2019
- Property-of-ceramics: technoceramics.co.in, Retrieved 17 May, 2019
- Structure-of-ceramic-materials: substech.com, Retrieved 19 April, 2019
- Ionic-and-covalent-bonding: substech.com, Retrieved 5 February, 2019
- Advantages-disadvantages-of-ceramics: hunker.com, Retrieved 26 July, 2019

2

Types of Ceramics

Some of the diverse types of ceramics are traditional ceramics, glass-ceramics, ceramic cements, structural ceramics, advanced ceramics, alumina ceramics, electroceramics and piezoelectric ceramics. This chapter has been carefully written to provide an easy understanding of these types of ceramics.

Traditional Ceramics

Traditional ceramics are ceramic materials that are derived from common, naturally occurring raw materials such as clay minerals and quartz sand. Through industrial processes that have been practiced in some form for centuries, these materials are made into such familiar products as china tableware, clay brick and tile, industrial abrasives and refractory linings, and Portland cement.

Traditional ceramic objects are almost as old as the human race. Naturally occurring abrasives were undoubtedly used to sharpen primitive wood and stone tools, and fragments of useful clay vessels have been found dating from the Neolithic Period, some 10,000 years ago. Not long after the first crude clay vessels were made, people learned how to make them stronger, harder, and less permeable to fluids by burning. These advances were followed by structural clay products, including brick and tile. Clay-based bricks, strengthened and toughened with fibres such as straw, were among the earliest composite materials. Artistic uses of pottery also achieved a high degree of sophistication, especially in China, the Middle East, and the Americas.

With the advent of the Metal Age some 5,000 years ago, early smiths capitalized on the refractory nature of common quartz sand to make molds for the casting of metals—a practice still employed in modern foundries. The Greeks and Romans developed lime-mortar cements, and the Romans in particular used the material to construct remarkable civil engineering works, some of which remain standing to this day. The Industrial Revolution of the 18th and 19th centuries saw rapid improvements in the processing of ceramics, and the 20th century saw a growth in the scientific understanding of these materials. Even in the age of modern advanced ceramics, traditional ceramic products, made in large quantities by efficient, inexpensive manufacturing methods, still make up the bulk of ceramics sales worldwide. The scale of plant operations can rival those found in the metallurgical and petrochemical industries.

Raw Materials

Because of the large volumes of product involved, traditional ceramics tend to be manufactured from naturally occurring raw materials. In most cases these materials are silicates—that is, compounds based on silica (SiO_2), an oxide form of the element silicon. In fact, so common is the use of silicate minerals that traditional ceramics are often referred to as silicate ceramics and their manufacture is often called the silicate industry. Many of the silicate materials are actually unmodified or chemically modified aluminosilicates (alumina [Al_2O_3] plus silica), although silica is also used in its pure form. Altogether, the raw materials employed in traditional ceramics fall into three commonly recognized groups: clay, silica, and feldspar.

Clay

Clay minerals such as kaolinite ($Al_2[Si_2O_5][OH]_4$) are secondary geologic deposits, having been formed by the weathering of igneous rocks under the influence of water, dissolved carbon dioxide, and organic acids. The largest deposits are believed to have formed when feldspar ($KAlSi_3O_8$) was eroded from rocks such as granite and was deposited in lake beds, where it was subsequently transformed into clay.

The importance of clay minerals to traditional ceramic development and processing cannot be overemphasized. In addition to being the primary source of aluminosilicates, these minerals have layered crystal structures that result in plate-shaped particles of extremely small micrometre size. When these particles are suspended in or mixed with water, the mixture exhibits unusual rheology, or flow under pressure. This behavior allows for such diverse processing methods as slip casting and plastic forming, which are described below. Clay minerals are therefore considered to be formers, allowing the mixed ingredients to be formed into the desired shape.

Silica and Feldspar

Other constituents of traditional ceramics are silica and feldspar. Silica is a major ingredient in refractories and white wares. It is usually added as quartz sand, sandstone, or flint pebbles. The role of silica is that of a filler, used to impart "green" (that is, unfired) strength to the shaped object and to maintain that shape during firing. It also improves final properties. Feldspars are aluminosilicates that contain sodium (Na), potassium (K), or calcium (Ca). They range in composition from $NaAlSi_3O_8$ and $KAlSi_3O_8$ to $CaAl_2Si_2O_8$. Feldspars act as fluxing agents to reduce the melting temperatures of the aluminosilicate phases.

Processing

Beneficiation

Compared with other manufacturing industries, far less mineral beneficiation (*e.g.*, washing, concentrating, and sizing of particulates) is employed for silicate ceramics. Clays going into common structural brick and tile are often processed directly as dug out of the ground, although there may be some blending, aging, and tempering for uniform distribution in water. Such impure clays are workable in untreated form because they already contain fillers and fluxes in association with the clay minerals. In the case of white wares, for which the raw materials must be in a purer state, the clays are washed, and impurities are either settled out or floated off. Silicas are purified by washing and separating unwanted minerals by gravity and by magnetic and electrostatic means. Feldspars are beneficiated by flotation separation, a process in which a frothing agent is added to separate the desired material from impurities.

Blending

The calculation of amounts, weighing, and initial blending of raw materials prior to forming operations is known as batching. Batching has always constituted much of the art of the ceramic technologist. Formulas are traditionally jealously guarded secrets, involving the selection of raw materials that confer the desired working characteristics and responses to firing and that yield the sought-after character and properties. Clays must be selected on the basis of workability, fusibility, fired color, and other requirements. Silicas, likewise, must meet criteria of chemical purity and particle size distribution.

Forming

The fine, platy morphology of clay particles is used to advantage in the forming of clay-based ceramic products. Depending upon the amount of water added, clay-water bodies can be stiff or plastic. Plasticity arises by virtue of the plate-shaped clay particles slipping over one another during flow. Nonclay ceramics can be similarly formed if plasticizers—usually polymers—are added to their mixes. In many cases organic binders are used to help hold the body together until it is fired. With even higher water content and the addition of dispersing agents to keep the clay particles in suspension, readily flowable suspensions can be produced. These suspensions are called slips or slurries and are employed in the slip casting of clay bodies.

Plastic Forming

Plastic forming is the primary means of shaping clay-based ceramics. After the raw materials are mixed and blended into a stiff mud or plastic mix, a variety of forming techniques are employed to produce useful shapes, depending upon the ceramic involved and the type of product desired. Foremost among these techniques are pressing and extrusion.

Pressing involves the application of pressure to eliminate porosity and achieve a specific shape, depending upon the die employed. Refractory bricks, for example, are often made by die presses that are either single-action (pressing from the top only) or dual-action (simultaneously pressing from top and bottom). Structural clay products such as brick and tile can be made in the same fashion. In pressing operations the feed material tends to have lower water content and is referred to as a stiff mud.

The problem with die casting is that it is a piecemeal rather than a continuous process, thereby limiting throughput. Many silicate ceramics are therefore manufactured by extrusion, a process that allows a more efficient continuous production. In a commercial screw-type extruder, a screw auger continuously forces the plastic feed material through an orifice or die, resulting in simple shapes such as cylindrical rods and pipes, rectangular solid and hollow bars, and long plates. These shapes can be cut upon extrusion into shorter pieces for bricks and tiles.

Slip Casting

A different approach to the forming of clay-based ceramics is taken in slip casting of white ware, as shown in figure. As mentioned above, with sufficient water content and the addition of suitable dispersing agents, clay-water mixtures can be made into suspensions called slurries or slips. These highly stable suspensions of clay particles in water arise from the careful manipulation of surface charges on the plate like clay particles. Without a dispersing agent, oppositely charged edges and surfaces of the particles would attract, leading to flocculation, a process in which groups of particles coagulate into flocs with a characteristic house-of-cards structure. Dispersing agents neutralize some of the surface charges, so that the particles can be made to repel one another and remain in suspension indefinitely. When the suspension is poured into a porous plaster mold, capillary forces suck the water into the mold from the slip and cause a steady deposition of clay particles, in dense face-to-face packing, on the inside surface of the mold. After a sufficient thickness of deposit has been obtained, the remaining slip can be poured off or drained and the mold opened to reveal a freestanding clay piece that can be dried and fired. Surprisingly complex shapes can be achieved through slip casting.

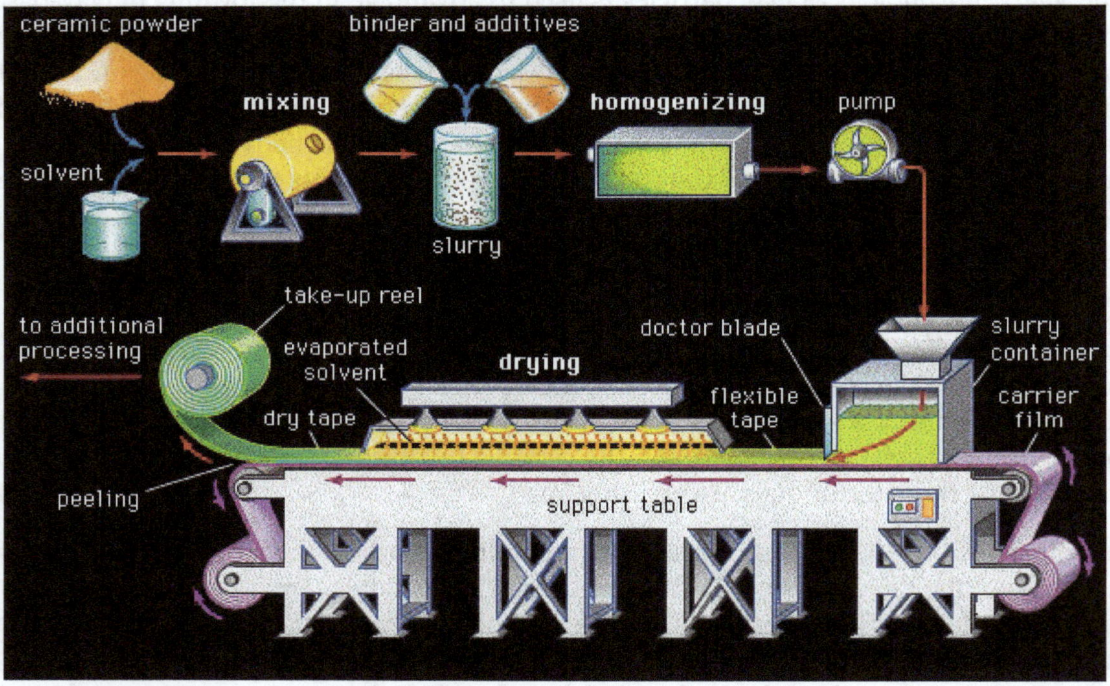

Firing

Kiln Operation

After careful drying to remove evaporable water, clay-based ceramics undergo gradual heating to remove structural water, to decompose and burn off any organic binders used in forming, and to achieve consolidation of the ware. Batches of specialty products, produced in smaller volumes, are cycled up and down in so-called batch furnaces. Most mass-produced traditional ceramics, on the other hand, are fired in tunnel kilns. These consist of continuous conveyor belt or railcar operations, with the ware traversing the kiln and gradually being heated from room temperature, through a hot zone, and back down to room temperature. Pyrometric cones, which deform and sag at specific temperatures, often ride with the ware to monitor the highest temperature seen in the traverse through the kiln.

Vitrification

The ultimate purpose of firing is to achieve some measure of bonding of the particles (for strength) and consolidation or reduction in porosity (*e.g.*, for impermeability to fluids). In silicate-based ceramics, bonding and consolidation are accomplished by partial vitrification. Vitrification is the formation of glass, accomplished in this case through the melting of crystalline silicate compounds into the amorphous, noncrystalline atomic structure associated with glass. As the formed ware is heated in the kiln, the clay component turns into progressively larger amounts of glass. The partial vitrification process can be analyzed through a phase diagram such as that shown in figure. In this diagram three crystalline phases are shown: the end members cristobalite (one crystallographic form of silica [SiO_2]) and alumina (Al_2O_3) and an intermediate compound, mullite ($3Al_2O_3 \cdot 2SiO_2$). The melting points of alumina and cristobalite, as shown on the left and right edges of the diagram, are quite high. However, intermediate compositions begin to melt at lower temperatures. As shown by the two horizontal lines on the diagram, melting begins to occur at 1,828 °C (3,322 °F) for high alumina compositions and as low as 1,587 °C (2,889 °F) for high silica compositions. These temperatures can be lowered still further by the addition of fluxing agents, such as alkali or alkaline-earth oxide feldspars. Between the two horizontal lines and the region of the diagram marked liquid, all compositions are only partly liquid (*e.g.*, mullite and liquid, alumina and liquid). This partial vitrification allows for the retention of solid particles, which helps to maintain the rigidity of the ceramic piece during firing in order to minimize sagging or warpage.

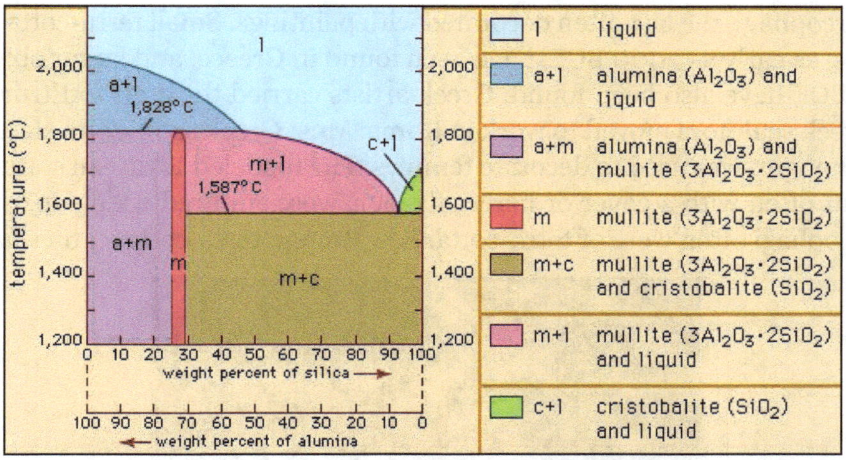

Phase diagram of the alumina-silica system. Depending on the temperature and on the content of silica and alumina, aluminosilicate clays, upon heating, form various combinations of alumina, cristobalite, mullite and liquid.

The role of the glassy liquid phase in the consolidation of fired clay objects is to facilitate liquid-phase or reactive-liquid sintering. In these processes the liquid first brings about a denser rearrangement of particles by viscous flow. Second, through solution-precipitation of the solid phases, small particles and surfaces of larger particles dissolve and reprecipitate at the growing "necks" that connect large particles. Rearrangement and solution-precipitation lead to bond formation and to progressive densification with reduction of porosity. A range of glass contents and residual porosities can be obtained, depending on the ingredients and the time the object is held at maximum temperature.

Finishing

If fired ceramic ware is porous and fluid impermeability is desired, or if a purely decorative finish is desired, the product can be glazed. In glazing, a glass-forming formulation is pulverized and suspended in an appropriate solvent. The fired ceramic body is dipped in or painted with the glazing slurry, and it is refired at a temperature that is lower than its initial firing temperature but high enough to vitrify the glaze formulation. Glazes can be colored by the addition of specific transition-metal or rare-earth elements to the glaze glass or by the suspension of finely divided ceramic particles in the glaze.

Terra-cotta

Terra-cotta is any kind of fired clay but, in general usage, a kind of object — e.g., vessel, figure, or structural form—made from fairly coarse, porous clay that when fired assumes a color ranging from dull ochre to red and usually is left unglazed. Most terra-cotta has been of a utilitarian kind because of its cheapness, versatility, and durability. Limitations in the basic materials often cause a superficial similarity between simply made works as far separated by time and distance as early Greece and the modern cultures of Latin America.

Throughout the ancient world, one of the most common uses of terra-cotta was for building-brick, roof tiles, and sarcophagi, the last often decorated with paintings. Small terra-cotta figures from the Early Bronze Age, as early as 3000 BCE, have been found in Greece, and larger objects dating from the 7th century BCE have also been found. Greek artists carried the craft to Etruria, whence both Etruscan and Greek sculptors moved to work in Rome. Most Greek terra-cotta statuary, more common than once thought, was used to decorate temples. The modeled Etruscan statues, at times very Greek in style but often with a gayer or fiercer flavour, were admired widely in antiquity. Figures on Etruscan sarcophagi often were of terra-cotta. Few Roman terra-cotta statues have been found.

Qin tomb: Life-size terra-cotta figures at Qin tomb, near Xi'an, Shaanxi province, China.

Molded statuettes 6 to 7 inches (15 to 18 cm) in height are common throughout the ancient world, among them very early primitive figures from Cyprus and painted, glazed human figures from Minoan Crete. The Cypriot figures often comprise groups of dancers or warriors, and the Cretan feature lively poses of women, horsemen, or animals. After the 7th century BCE, styles became less hieratic, the subjects more mundane—e.g., a nurse with child, a teacher and pupils, an actor in costume. The Tanagra figurines, found in Tanagra in central Greece (Boeotia), are the best known of this kind. In the Hellenistic period, from the 4th century BCE, centres of statuette production moved to Asia Minor and westward, being found throughout the Roman Empire as far as Britain. Styles in the East became more ornate and influenced by Oriental values in design and subject.

Architectural relief, especially where wood or clay was used for building, used floral or more abstract designs and such figured representations as chariot races or animal or female heads; examples have been found in Asia Minor, Greece, and Etruscanized southern Italy. Votive reliefs also were common, notably those of the local divinities and heroes rendered broadly and smoothly in Tarentum (Taranto), southern Italy, and the small, meticulous reliefs of local cults found at Locris in central Greece. The fine 5th-century reliefs from the island of Melos, in which mythological scenes predominate, decorated chests. Much Roman architecture is decorated with relief themes from mythology, especially of Dionysus and his revellers.

The use of terra-cotta for all purposes virtually died out between the end of the Roman Empire and the 14th century. In 15th-century Italy and Germany it appeared again, either molded or carved, and in its natural color as friezes, moldings, or inset medallions decorating buildings. A new use of terra-cotta was in the highly glazed and colored sculpture introduced in Florence early in the 15th century by the Della Robbia family. The effect, adding a freshness of accent especially to marble and stone, was imitated widely, and the use of terra-cotta, glazed or unglazed, spread throughout Europe. Free sculpture in terra-cotta also was revived in the 15th century by such artists as Donatello, Verrocchio, and especially Guido Mazzoni and Antonio Begarelli working in Modena; often it was painted in natural colors or to imitate marble or bronze.

Virgin and Child, glazed terra-cotta from the workshop of Benedetto Buglioni in a gilt wood Renaissance Revival frame.

During the following centuries, most terra-cotta figures were executed as preliminary studies, though the works of such 18th-century French artists as Jean-Baptiste Lemoyne and Jean-Antoine Houdon display a personal immediacy of subject that is not transferable to the harder material. In the same period, such pottery centres as Sèvres in France introduced finely wrought small groups with allegorical and mythological themes. Terra-cotta was used both architecturally and for figures during the 19th century, but its modern revival dates from the 20th century, when both potters and architects again became interested in the aesthetic properties of the material.

Porcelain

The term porcelain refers to a wide range of ceramic products that have been baked at high temperatures to achieve vitreous, or glassy, qualities such as translucence and low porosity. Among the most familiar porcelain goods are table and decorative china, chemical ware, dental crowns, and electrical insulators. Usually white or off-white, porcelain comes in both glazed and unglazed varieties, with bisque, fired at a high temperature, representing the most popular unglazed variety.

Although porcelain is frequently used as a synonym for china, the two are not identical. They resemble one another in that both are vitreous wares of extremely low porosity, and both can be glazed or unglazed. However, china, also known as soft-paste or tender porcelain, is softer: it can be cut with a file, while porcelain cannot. This difference is due to the higher temperatures at which true porcelain is fired, 2,650 degrees Fahrenheit (1,454 degrees Celsius) compared to 2,200 degrees Fahrenheit (1,204 degrees Celsius) for china. Due to its greater hardness, porcelain has some medical and industrial applications which china, limited to domestic and artistic use, does not. Moreover, whereas porcelain is always translucent, china is opaque.

Hard-paste or "true" porcelain originated in China during the T'ang dynasty (618-907 A.D.); however, high quality porcelain comparable to modern wares did not develop until the Yuan dynasty (1279-1368 A.D.). Early Chinese porcelain consisted of kaolin (china clay) and pegmatite, a coarse type of granite. Porcelain was unknown to European potters prior to the importation of Chinese wares during the Middle Ages. Europeans tried to duplicate Chinese porcelain, but, unable to analyze its chemical composition, they could imitate only its appearance. After mixing glass with tin oxide to render it opaque, European craftspeople tried combining clay and ground glass. These alternatives became known as soft-paste, glassy, or artificial porcelains. However, because they were softer than genuine porcelain, as well as expensive to produce, efforts to develop true porcelain continued. In 1707 two Germans named Ehrenfried Walter von Tschimhaus and Johann Friedrich Bottger succeeded by combining clay with ground feldspar instead of the ground glass previously used.

Later in the eighteenth century the English further improved upon the recipe for porcelain when they invented bone china by adding ash from cattle bones to clay, feldspar, and quartz. Although bone china is fired at lower temperatures than true porcelain, the bone ash enables it to become translucent nonetheless. Because it is also easier to make, harder to chip, and stronger than hard porcelain, bone china has become the most popular type of porcelain in the United States and Britain (European consumers continue to favor hard porcelain).

Raw Materials

The primary components of porcelain are clays, feldspar or flint, and silica, all characterized by small particle size. To create different types of porcelain, craftspeople combine these raw materials in varying proportions until they obtain the desired green (unfired) and fired properties.

Figure indicates to make porcelain, the raw materials—such as clay, felspar, and silica—are first crushed using jaw crushers, hammer mills, and ball mills. After cleaning to remove improperly sized materials, the mixture is subjected to one of four forming processes—soft plastic forming, stiff plastic forming, pressing, or casting—depending on the type of ware being produced. The ware then undergoes a preliminary firing step, bisque-firing.

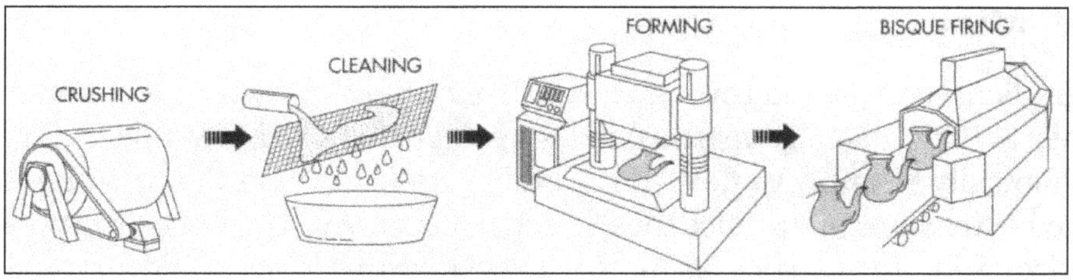

Although the composition of clay varies depending upon where it is extracted and how it is treated, all clays vitrify (develop glassy qualities), only at extremely high temperatures unless they are mixed with materials whose vitrification threshold is lower. Unlike glass, however, clay is refractory, meaning that it holds its shape when it is heated. In effect, porcelain combines glass's low porosity with clay's ability to retain its shape when heated, making it both easy to form and ideal for domestic use. The principal clays used to make porcelain are china clay and ball clay, which consist mostly of kaolinate, a hydrous aluminum silicate.

Feldspar, a mineral comprising mostly aluminum silicate, and flint, a type of hard quartz, function as fluxes in the porcelain body or mixture. Fluxes reduce the temperature at which liquid glass forms during firing to between 1,835 and 2,375 degrees Fahrenheit (1,000 and 1,300 degrees Celsius). This liquid phase binds the grains of the body together.

Silica is a compound of oxygen and silicon, the two most abundant elements in the earth's crust. Its resemblance to glass is visible in quartz (its crystalline form), opal (its amorphous form), and sand (its impure form). Silica is the most common filler used to facilitate forming and firing of the body, as well as to improve the properties of the finished product. Porcelain may also contain alumina, a compound of aluminum and oxygen, or low-alkali containing bodies, such as steatite, better known as soapstone.

Manufacturing Process

After the raw materials are selected and the desired amounts weighed, they go through a series of preparation steps. First, they are crushed and purified. Next, they are mixed together before being subjected to one of four forming processes—soft plastic forming, stiff plastic forming, pressing, or casting; the choice depends upon the type of ware being produced. After the porcelain has been formed, it is subjected to a final purification process, bisque-firing, before being glazed. Glaze is a layer of decorative glass applied to and fired onto a ceramic body. The final manufacturing phase is firing, a heating step that takes place in a type of oven called a kiln.

Crushing the Raw Materials

- First, the raw material particles are reduced to the desired size, which involves using a variety of equipment during several crushing and grinding steps. Primary crushing is done in jaw crushers who use swinging metal jaws. Secondary crushing reduces particles to 0.1 inch (0.25 centimeter) or less in diameter by using mullers (steel-tired wheels) or hammer mills, rapidly moving steel hammers. For fine grinding, craftspeople use ball mills that consist of large rotating cylinders partially filled with steel or ceramic grinding media of spherical shape.

Cleaning and Mixing

- The ingredients are passed through a series of screens to remove any under- or over-sized materials. Screens, usually operated in a sloped position, are vibrated mechanically or electromechanically to improve flow. If the body is to be formed wet, the ingredients are then combined with water to produce the desired consistency. Magnetic filtration is then used to remove iron from the slurries, as these watery mixtures of insoluble material are called. Because iron occurs so pervasively in most clays and will impart an undesirable reddish hue to the body if it oxidizes, removing it prior to firing is essential. If the body is to be formed dry, shell mixers, ribbon mixers, or intensive mixers are typically used.

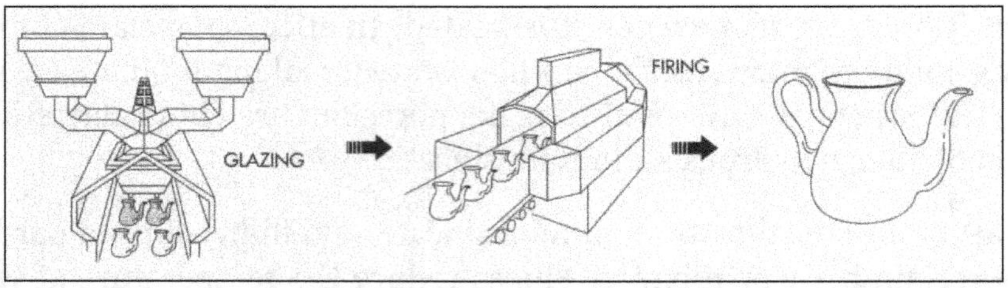

After bisque firing, the porcelain wares are put through a glazing operation, which applies the proper coating. The glaze can be applied by painting, dipping, pouring, or spraying. Finally, the ware undergoes a firing step in an oven or kiln. After cooling, the porcelain ware is complete.

Forming the Body

- Next, the body of the porcelain is formed. This can be done using one of four methods, depending on the type of ware being produced:
    - Soft plastic forming, where the clay is shaped by manual molding, wheel throwing, jiggering, or ram pressing. In wheel throwing, a potter places the desired amount of body on a wheel and shapes it while the wheel turns. In jiggering, the clay is put on a horizontal plaster mold of the desired shape; that mold shapes one side of the clay, while a heated die is brought down from above to shape the other side. In ram pressing, the clay is put between two plaster molds, which shape it while forcing the water out. The mold is then separated by applying vacuum to the upper half of the mold and pressure to the lower half of the mold. Pressure is then applied to the upper half to free the formed body.
    - Stiff plastic forming, which is used to shape less plastic bodies. The body is forced through a steel die to produce a column of uniform girth. This is either cut into the desired length or used as a blank for other forming operations.
    - Pressing, this is used to compact and shape dry bodies in a rigid die or flexible mold. There are several types of pressing, based on the direction of pressure. Uniaxial pressing describes the process of applying pressure from only one direction, whereas isostatic pressing entails applying pressure equally from all sides.
    - Slip casting, in which slurry is poured into a porous mold. The liquid is filtered out through the mold, leaving a layer of solid porcelain body. Water continues to drain out of the cast layer, until the layer becomes rigid and can be removed from the mold. If the

excess fluid is not drained from the mold and the entire material is allowed to solidify, the process is known as solid casting.

Bisque-firing

After being formed, the porcelain parts are generally bisque-fired, which entails heating them at a relatively low temperature to vaporize volatile contaminants and minimize shrinkage during firing.

Glazing

After the raw materials for the glaze have been ground they are mixed with water. Like the body slurry, the glaze slurry is screened and passed through magnetic filters to remove contaminants. It is then applied to the ware by means of painting, pouring, dipping, or spraying. Different types of glazes can be produced by varying the proportions of the constituent ingredients, such as alumina, silica, and calcia. For example, increasing the alumina and decreasing the silica produces a matte glaze.

Firing

- Firing is a further heating step that can be done in one of two types of oven, or kiln. A periodic kiln consists of a single, refractory-lined, sealed chamber with burner ports and flues (or electric heating elements). It can fire only one batch of ware at a time, but it is more flexible since the firing cycle can be adjusted for each product. A tunnel kiln is a refractory chamber several hundred feet or more in length. It maintains certain temperature zones continuously, with the ware being pushed from one zone to another. Typically, the ware will enter a preheating zone and move through a central firing zone before leaving the kiln via a cooling zone. This type of kiln is usually more economical and energy efficient than a periodic kiln.
- During the firing process, a variety of reactions take place. First, carbon-based impurities burn out, chemical water evolves (at 215 to 395 degrees Fahrenheit or 100 to 200 degrees Celsius), and carbonates and sulfates begin to decompose (at 755 to 1,295 degrees Fahrenheit or 400 to 700 degrees Celsius). Gases are produced that must escape from the ware. On further heating, some of the minerals break down into other phases, and the fluxes present (feldspar and flint) react with the decomposing minerals to form liquid glasses (at 1,295 to 2,015 degrees Fahrenheit or 700 to 1,100 degrees Celsius). These glass phases are necessary for shrinking and bonding the grains. After the desired density is achieved (greater than 2,195 degrees Fahrenheit or 1,200 degrees Celsius), the ware is cooled, which causes the liquid glass to solidify, thereby forming a strong bond between the remaining crystalline grains. After cooling, the porcelain is complete.

Quality Control

The character of the raw materials is important in maintaining quality during the manufacturing process. The chemical composition, mineral phase, particle size distribution, and colloidal surface area affect the fired and unfired properties of the porcelain. With unfired body, the properties evaluated include viscosity, plasticity, shrinkage, and strength. With fired porcelain, strength, porosity, color, and thermal expansion are measured. Many of these properties are monitored and controlled during manufacturing using statistical methods. Both the raw materials and the process parameters (milling time and forming pressure, for example) can be adjusted to achieve desired quality.

High-quality porcelain art and dinnerware will continue to enhance the culture. Improvements in manufacturing will continue to increase both productivity and energy efficiency. For instance, a German kiln manufacturer has developed a prefabricated tunnel kiln for fast firing high-quality porcelain in less than 5 hours. Firing is achieved by partly reducing atmosphere at a maximum firing temperature of 2,555 degrees Fahrenheit (1,400 degrees Celsius). The kiln uses high-velocity burners and an automatic control system, producing 23,000 pounds (11,500 kilograms) of porcelain in 24 hours.

Manufacturers of porcelain products may also have to increase their recycling efforts, due to the increase in environmental regulations. Though unfired scrap is easily recycled, fired scrap poses a problem: mechanically strong and therefore hard to break down, it is usually dumped into landfills. However, preliminary research has shown that fired scrap can be reused after thermal quenching (where the scrap is reheated and then quickly cooled), which makes it weaker and easier to break down. The scrap can then be used as a raw material. Porcelain appears to be playing a more important role in technical applications.

Whiteware

Whiteware is any of a broad class of ceramic products that are white to off-white in appearance and frequently contain a significant vitreous, or glassy, component. Including products as diverse as fine china dinnerware, lavatory sinks and toilets, dental implants, and spark-plug insulators, whitewares all depend for their utility upon a relatively small set of properties: imperviousness to fluids, low conductivity of electricity, chemical inertness, and an ability to be formed into complex shapes. These properties are determined by the mixture of raw materials chosen for the products, as well as by the forming and firing processes employed in their manufacture.

Raw Materials: Clay, Flint and Feldspar

Whitewares are often referred to as triaxial bodies, owing to the three mineral types—clay, silica, and feldspar—consistently found in their makeup. Clay is the plastic component, giving shaping abilities to the unfired product and also serving as a glass former during firing. Flint (the common name used in the industry for all forms of silica) serves as filler, lending strength to the shaped body before and during firing. Feldspar serves as a fluxing agent, lowering the melting temperatures of the mixture.

Clay is the most important of the ingredients, and the most important clay used in fine whiteware products is kaolin, also known as china clay. Kaolin is the only type of clay from which a white, translucent, vitreous ceramic can be made. It is refractory clay, meaning that it can be fired at high temperatures without deforming, and it is white-burning, meaning that it imparts whiteness to the finished ware. Kaolin is formed principally of the mineral kaolinite, a hydrous aluminosilicate with a fine, platy structure; its ideal chemical formula is $Al_2(Si_2O_5)(OH)_4$. China clays are composed mostly of well-ordered kaolinite, with no impurities. Lower-grade whitewares are usually made of ball clays, which incorporate ordered and disordered kaolinite plus other clay minerals and impurities. These impurities—particularly iron oxides—render the fired ware off-white to gray or tan in color.

Products

Whiteware products are often differentiated into three main classes—porous, semivitreous, and vitreous—according to their degree of vitrification (and resulting porosity). Proceeding from porous to vitreous, more particular product categories include earthenware, stoneware, china, and technical porcelains. Earthenware is nonvitreous and of medium porosity. It is often glazed to provide fluid impermeability and an attractive finish. Specific products include tableware and decorative tile ware. Stoneware is a semivitreous or vitreous whiteware with a fine microstructure (that is, a fine arrangement of solid phases and glass on the micrometre level). Products include tableware, cookware, chemical ware, and sanitary ware (e.g., drainpipe).

All vitreous whitewares are often referred to as porcelains, but in the ceramics industry a distinction is maintained between the true porcelains (or technical porcelains) and china. China is vitreous whiteware for nontechnical applications. Because of its high glass content, it can be used unglazed, though it also can be glazed for aesthetic appeal. China is known for high strength and impact resistance and also for low water absorption—all deriving from the high glass content. Typical products include hotel china, a lower grade of china tableware with a strength and impact resistance suiting it to commercial use; fine china (including bone china), a highly vitreous, translucent tableware; and sanitary plumbing fixtures.

Technical porcelains, like china, are vitreous and nonporous. They are similarly strong and impact-resistant, but they are also chemically inert in corrosive environments and are excellent insulators against electricity. Applications include chemical ware, dental implants, and electric insulators, including spark-plug insulators in automobile engines.

Processing

Typically, pressing is employed in the forming of tiles, chemical ware, and technical porcelains; extrusion in the forming of tiles and sanitary ware (including pipe); and slip casting in the forming of plumbing fixtures and some tableware. In addition to these standard processes, jiggering is employed in the manufacture of tableware. Jiggering involves the mixing of a plastic mass and turning it on a wheel beneath a template to a specified size and shape.

Figure shows stages in the slip casting of a thin-walled whiteware container. Clay powder is mixed in water together with a dispersing agent, which keeps the clay particles suspended evenly throughout the clay-water slurry, or slip. The slip is poured into a plaster mold, where water is drawn out by capillary action and a cast is formed by the deposition of clay particles on the inner surfaces of the mold. The remaining slip is drained, and the cast is allowed to dry partially before the drain hole is plugged and the mold separated. The unfinished ware is given a final drying in an oven before it is fired into a finished product.

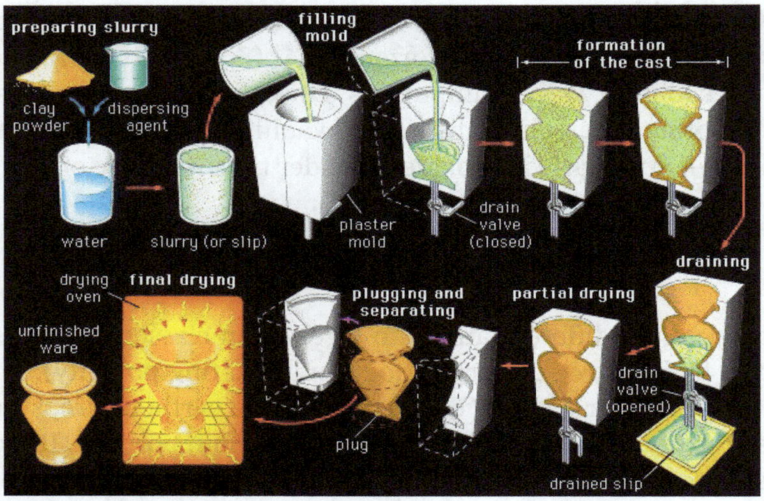

Most whitewares are fired in continuous tunnel kilns. The porous varieties are fired at lower temperatures (1,100–1,250 °C, or approximately 2,000–2,300 °F), whereas china and true porcelains are fired at 1,250 to 1,300 °C (2,300 to 2,400 °F). Porous and semivitreous whitewares may be glazed in a second firing to produce an impermeable glass coating for decorative or functional purposes.

Unfired whiteware stacked on kiln carts prior to firing.

One of the great advantages of the triaxial composition of whitewares is that it makes the formed piece relatively insensitive to minor changes in composition and in firing time or temperature. This stability is a result of the wide range of temperatures over which the three ingredients melt to form glass. As an example, in a typical feldspar-clay-silica composition for porcelain, a whiteware with a particularly high glassy component, small grains of feldspar would begin to form liquid at temperatures as low as 990 °C (1,810 °F), and large feldspar grains would be molten by 1,140 °C (2,080 °F). Because of the high viscosity of the liquid formed, there would be no change in the shape of the ceramic piece until approximately 1,200 °C (2,200 °F). Above this temperature the feldspar grains would react with surrounding clay particles to form glass, and “needles” of mullite (a crystalline aluminosilicate mineral formed during the firing of clay-silica mixtures) would grow into the liquid regions. In addition, the surfaces of silica particles would begin to dissolve and form solution rims, or envelopes of glass surrounding the crystalline particle. As more and more of the silica particles dissolved, the resulting glass would become increasingly viscous, helping to maintain the integrity of the piece.

Silicate Ceramics

Silicate ceramics have clearly defined thermal properties and, as such, are particularly useful for advanced engineering and technology.

The main types of silicate ceramics are based either on alumosilicates (kaolin or claybased ceramic such as porcelain, earthenware, stoneware and bricks; system K_2O-Al_2O_3-SiO_2) or on magnesium silicates (talc-based technical ceramics such as steatite, cordierite and forsterite ceramics; system MgO-Al_2O_3-SiO_2). Special groups are zircon- and mullitebased fine ceramics (for electrical insulators) as well as low-thermal expansion ceramics in the system Li_2O-Al_2O_3-SiO_2 (for thermoshock-resistant tableware), which have an extremely narrow sintering interval and are therefore preferentially produced by glass-ceramic techniques (i.e. devitrification of glasses by nucleation and growth). Silicate ceramics are conventionally subdivided into coarse or fine and according to water absorption, into dense (< 2% for fine and < 6% for coarse) or porous ceramics (> 2% and > 6%, respectively).

Kaolin and Clay-based Ceramics and Dental Porcelain

In the ternary raw material diagram (kaolin/clay–feldspar–quartz) the composition of all kaolin- and clay-based ceramics lies in the mullite field, that of dental porcelain in the leucite field.

- Main types of dense silicate fine ceramics:
    - Hard porcelain (the majority of middle-European porcelains – common firing temperatures approx. 1400 ± 50 °C).
    - Soft porcelain (e.g. old Asian and "vitreous china", firing temperature 1200– 1300 °C, higher content of fluxes).
    - Bone china (English, based on up to 50% bone ash as a raw material – apart from kaolin, quartz), frit porcelain (French, based on glass frits, 1150 °C) and "parian" (unglazed ornamental porcelain with a low kaolin content of < 40%).
    - Dental porcelain (feldspar content approx. 80%, kaolin < 5%, firing temperature < 1250 °C).
    - Electrotechnical porcelain for insulators; usually containing alumina (instead of quartz in earlier types of electrotechnical porcelain) in order to increase the mechanical and electrical strength (firing temperature approx. 1250 °C).
- Hard porcelain:
    - Porcelain is a densely sintered (defined by a water absorption < 2%), white and translucent fine ceramic material prepared from natural raw materials; typical raw

material mixture for hard porcelain: 50% kaolin (part of which can be replaced by plastic clay), 25% quartz and 25% feldspar (preferentially K-feldspar).

- Hard porcelain: after firing at 1400 ± 50 °C (with a relatively broad sintering interval) the final ceramic body consists of minimum 50% glass phase, maximum 25% mullite and maximum 25% residual quartz (which can be partly transformed to cristobalite); typical properties: density 2.3–2.5 g/cm³, tensile modulus 70–80 GPa, Poisson ratio 0.17, flexural strength up to 100 MPa, thermal conductivity 1.2–1.6 W/mK, thermal expansion coefficient $4–6 \cdot 10^{-6}$ K^{-1}.
- At temperatures < 1100 °C the clay minerals (mainly kaolinite) dehydrate (metakaolinite above 500–600 °C) and form transient phases by releasing silica (defective spinel phase above 900–1000 °C), quartz undergoes polymorphic transitions and mixed K-Na-feldspars (perthites) may homogenize; at temperatures > 1100 °C: formation of feldspar melt causing liquid phase sintering (vitrification), partial dissolution of quartz (viscosity increase) and formation of mullite, either directly from the clay minerals (primary mullite) or by reaction of the clay minerals with the feldspar melt (secondary mullite).

- Other types of kaolin- and clay-based silicate ceramics:
  - Earthenware: Porous, non-transparent fine ceramics with a white or colored body; typically fired at 1200 ± 50 °C and glazed with a PbO-containing glaze in a second firing cycle (at approx. 1100 °C); typical raw material compositions are 50–55% clay, 40 ± 5% quartz and 5–10% feldspar; commonly used for tableware and tiles (for the latter, however, firing is a single-step process at about 1100 °C); "faience" is earthenware with a white body, "majolika" with a colored body, "terracotta" is coarse earthenware.
  - Stoneware: Porous, non-transparent coarse ceramics with a colored body, typically fired at 1250 ± 50 °C (for sanitary ware, floor tiles and sewer pipes with a brown NaCl glaze); note, however, that another variant of stoneware (for tableware, tiles and chemical vessels) is a dense fine ceramic (vitrified stoneware); sanitary ware is between porcelain and stoneware.
  - Bricks: Porous coarse ceramics, produced from cheap, local clays and loams, typically fired at 900–1000 °C; the loams should not contain pyrite and sulfates → $CaSO_4$ near the body surface (→ hydratation → volume expansion), neither calcite ($CaCO_3$) → may remain unreacted as free CaO after firing (→ hydratation → volume expansion); important properties: frost resistance (requires low porosity) and thermal insulation (requires high porosity).

Talc-based Technical Ceramics

All talc-based technical ceramics (ternary phase diagram MgO-Al_2O_3-SiO_2) require precise firing control (narrow sintering interval of a few °C).

- Steatite ceramics: Basic raw materials – talc and clay (+ feldspar or $BaCO_3$); desired phase – protoenstatite in approx. 30% glassy matrix; firing temperature 1350–1370 °C; problem to be controlled: transformation of the high-temperature proto- into (low temperature) clinoenstatite, accompanied with volume expansion.

- Cordierite ceramics: Basic raw materials – talc, clay and Al_2O_3; desired phase – cordierite in a glassy matrix; self-glazing effect by non-wetting melt exuded onto the surface; low thermal expansion coefficient ($2{\cdot}10^{-6}$ K^{-1}). Note that only silica glass and Li_2O-Al_2O_3-SiO_2 (glass) ceramics exhibit lower thermal expansion coefficients ($< 0.5{\cdot}10^{-6}$ K^{-1}) higher thermal shock resistance.

- Forsterite ceramics: Basic raw materials – talc and clay (+ $MgCO_3$); firing is less sensitive with respect to temperature (since at 1360 °C only a small amount of eutectic melt is formed and this amount does not change very much with temperature), but very sensitive to changes in composition; high thermal expansion coefficient ($11{\cdot}10^{-6}$ K^{-1}) enables welding with metals (vaccuum electrotechnics).

Complex exercise problem: Use the ternary phase diagrams of the systems K_2O-Al_2O_3-SiO_2 and MgO-Al_2O_3-SiO_2 to discuss the phase composition of hard porcelain, cordierite ceramics, forsterite ceramics and steatite ceramics.

Magnesium Silicate Ceramics

Derived from soapstone, clay, corundum and mullite, the chemical composition of magnesium silicate ceramics includes silica (SiO_2), magnesia (MgO) and some traces of alumina (Al_2O_3). Steatite and cordierite are two types of magnesium silicate ceramics that share a similar chemical structure, but vary on their density and melting points, which subsequently affects the different ways in which these silicate ceramics are often utilized. For example, steatite has a density of 3.21 g/cm^3 and a melting point of 1557 °C. Steatite ceramics often exhibit a high mechanical strength, dielectric properties and a low loss factor, thereby allowing this material to typically be utilized in heat and electrical engineering applications. Such consumer products include electrical sockets, control housings, insulating beads, low-voltage power fuses and base plates.

In contrast, both the density and melting point of cordierite ceramics is slightly lower than that of steatite, at 2.60 g/cm^3 and 1470 °C, respectively. As a result, cordierite ceramics often exhibit superior properties when compared to steatite ceramics, including excellent thermal shock resistance, high chemical durability, high refractoriness and a low thermal expansion coefficient (CTE) that typically measures to 1.5×10^{-6} C^{-1} when present in a temperature range of 25-700 °C, whereas steatite ceramics have a much higher CTE value of 6×10^{-6} C^{-1} when present in the same external conditions. Some applications of cordierite ceramics include electrical porcelains, catalytic convertor substrates that are used to control the production of gas exhaust in vehicles, heat exchangers in gas turbine engines, industrial furnaces, metal refractory coatings, integrated circuit substrates and much more.

Cordierite ceramics can be manufactured a much lower cost than steatite, while simultaneously providing better electrical properties than its counterpart; making this type of magnesium silicate ceramic a popular material option in the electronics industry.

Leading Producers of Magnesium Silicate Ceramics

There are five applications in which the magnesium silicate ceramics market divides these materials and their products into, which include:

1. Electronics and Electricals,
2. Transportation,
3. Medical,
4. Industrial Market,
5. Defense and Security.

Glass-ceramics

Glass-ceramics are fine-grained polycrystalline materials formed when glasses of suitable compositions are heat treated and thus undergo controlled crystallisation to the lower energy, crystalline state. It must be emphasised here that only specific glass compositions are suitable precursors for glass-ceramics due to the fact that some glasses are too stable and difficult to crystallise whereas others result in undesirable microstructures by crystallising too readily in an uncontrollable manner. In addition, it must also be accentuated that in order for a suitable product to be attained, the heat-treatment is critical for the process and a range of generic heat treatment procedures are used which are meticulously developed and modified for a specific glass composition.

A glass-ceramic is formed by the heat treatment of glass which results in crystallisation. Crystallisation of glasses is attributed to thermodynamic drives for reducing the Gibbs' free energy, and the Amorphous Phase Separation (APS) which favours the crystallisation process by forming a nucleated phase easier than it would in the original glass. When a glass is melted, the liquid formed from the melting might spontaneously separate into two very viscous liquids or phases. By cooling the melt to a temperature below the glass transformation region it will result in the glass being phase separated and this is called liquid-liquid immiscibility. This occurs when both the phases are liquid. Hence a glass can simply be considered as a liquid which undergoes a demixing process when it cools. The immiscibility is either stable or metastable depending on whether the phase seperation occurs above or below the liquidus temperature respectively. The metastable immiscibility is much more inmportant and has two processes which then cause phase seperation and hence crystallisation; nucleation and crystal growth and spinodal decomposition.

The first APS process has two distinguished stages; Nucleation (whereby the crystals will grow to a detectable size on the nucleus) and Crystal growth. Nucleation can either be homogeneous; where the crystals form spontaneously within the melt or heterogeneous; crystals form at a pre-existing surface such as that due to an impurity, crucible wall etc. Many a time the parent glass composition is specifically chosen to contain species which enhance internal nucleation which in the majority of cases is required. Such species also called nucleating agents can include metallic agents such as Ag, Pt and Pd or non-metallic agents such as TiO_2, P_2O_5 and fluorides. The second process is spinodal decomposition which involves a gradual change in composition of the two phases until they reach the immiscibility boundary. As both the processes for APS are different, the glass formed will clearly result in having different morphology to each other.

A glass-ceramic is usually not fully crystalline; with the microstructure being 50-95 volume % crystalline with the remainder being residual glass. When the glass undergoes heat treatment, one or more crystalline phases may form. Both the compositions of the crystalline and residual glass are different to the parent glass. In order for glass-ceramics having desirable properties to be developed, it is crucial to control the crystallisation process so that an even distribution of crystals can be formed. This is done by controlling the nucleation and crystal growth rate. The nucleation rate and crystal growth rate is a function of temperature and are accurately measured experimentally.

The aim of the crystallisation process is to convert the glass into glass-ceramic which have properties superior to the parent glass. The glass-ceramic formed depends on efficient internal nucleation from controlled crystallisation which allows the development of fine, randomly oriented grains without voids, microcracks, or other porosity. This results in the glass-ceramic being much stronger, harder and more chemically stable than the parent glass.

Glass-ceramics are characterised in terms of composition and microstructure as their properties depend on both of these. The ability of a glass to be formed as well as its degree of workability depends on the bulk composition which also determines the grouping of crystalline phases which consecutively govern the general physical and chemical characteristics, e.g. hardness, density, acid resistance, etc. As mentioned earlier, nucleating agents are used in order for internal nucleation to occur so that the glass-ceramic produced has desirable properties. Microstructure is the key to most mechanical and optical properties; it can promote or diminish the role of the key crystals in the glass-ceramic. The desirable properties obtained from glass-ceramics are crucial in order for them to have applications in the field of biomaterials.

Glass-ceramics are used as biomaterials in two different fields: First, they are used as highly durable materials in restorative dentistry and second, they are applied as bioactive materials for the replacement of hard tissue. Dental restorative materials are materials which restore the natural tooth structure (both in shape and function), exhibit durability in the oral environment, exhibit high strength and are wear resistance. In order for dental restorative materials to restore the natural tooth structure, it is crucial to maintain the vitality of the tooth. . However non-vital teeth may also be treated with restorative materials to reconstruct or preserve the aesthetic and functional properties of the tooth.

In order for glass-ceramics to be used for dental applications, they must possess high chemical durability, mechanical strength and toughness and should exhibit properties which mimic the natural tooth microstructure in order for it to be successful as an aesthetic. Glass-ceramics allow all these properties to be united within one material. As mentioned previously, for a glass-ceramic to have the desired properties, the glass is converted into a glass-ceramic via controlled crystallisation to achieve the crystal phase wanted and hence the desired properties it could possibly have. Hence, the glass-ceramic developed allows it to have properties such as low porosity, increased strength, durability, toughness etc which are crucial in the field of dental restorations as it prevents restorative failures which are mainly due to stress and porosity which causes cracks and hence failures.

It took many years of research in order to get a material strong enough to be initially used as a dental reconstructive material. However over the past 10-15 years, research has progressed vastly and now glass-ceramics demonstrate good strength, high durability and good aesthetics. The development and processing of glass-ceramics has been focused on particular clinical applications, such as dental inlays, crowns, veneers, bridges and dental posts with abutments.

Glass-ceramics are divided into seven types of materials:

1. Mica glass-ceramics,
2. Mica apatite glass-ceramics,
3. Leucite glass-ceramics,
4. Leucite apatite glass-ceramics,
5. Lithium Disilicate glass-ceramics,
6. Apatite containing glass-ceramics,
7. ZrO_2-containing glass-ceramics.

The first commercially usable glass ceramic products for restorative dentistry were composites of mica glass ceramics. Dicor and Dicor MGC were products based on these. According to the mechanism of controlled volume crystallisation of glasses, tetrasilicic micas, $Mg_{2.5}Si_4O_{10}F_2$, showing crystal sizes of 1 to 2 µm in the glass ceramic were produced. Dicor being amongst them was shaped by means of centrifugal casting methods to produce dental restorations such as dental crowns and inlays. Depending on the different crystal sizes and the corresponding microstructure of the glass ceramic, it was also possible to manufacture glass ceramics for machining applications. Dicor MGC being amongst them. This resulted in the characteristic of good machinability in this type of glass-ceramic to be exploited and results concluded that crystals upto only 2 µm in length in the material improved mechanical strength over other materials.

Mica-apatite glass-ceramics have been produced in the SiO_2-Al_2O_3-Na_2O-K_2O-MgO-CaO-P_2O_5-F system. The main crystal phases are phlogopite, $(K,Na)Mg_3(AlSi_3O_{10})F_2$ and fluorapatite, $Ca_5(PO_4)_3F$. The base glass consists of three glass phases: a large droplet-shaped phosphate-rich phase, a small droplet-shaped silicate and a silicate glass matrix. Mica is formed during heat treatment, as in apatite-free glass-ceramics, by in-situ crystallization via the mechanism of volume crystallization. Apatite is formed within the phosphate-rich droplet phase. Astonishingly, every single apatite crystal possesses its own nucleation site in the form of a single phosphate drop. The glass-ceramic is biocompatible and suitable for applications in head and neck surgery as well as in the field of orthopaedics.

Leucite glass-ceramics can be formed by applying the advantage of the viscous flow mechanism. IPS Empress is of this type of glass-ceramic. The material is processed by using the lost wax technique, whereby a wax pattern of the dental restoration such as an inlay, onlay, veneer or crown is produced and then put in a refractory die material. Then the wax is burnt out to create space to be filled by the glass-ceramic. As the glass-ceramic has a certain volume of glass phase, the principle of viscous flow can be applied and hence the material can be pressed into a mould. Surface crystallisation and surface nucleation mechanisms were controlled in order for this type of glass-ceramic to be formed. Consequently, the manufacturing of inlays and crowns developed due to the application of viscous flow mechanism of glass-ceramics in different shapes. The resulting leucite glass-ceramic restorations transluceny, color and wear resistance behavior can then be adjusted to those of natural tooth. Additionally, the leucite glass-ceramic restorations can be produced by machining with CAD/CAM. IPS ProCAD and IPS Empress CAD are glass ceramics produced via this method. All leucite glass-ceramic restorations are bonded to the tooth structure with a luting material, preferably an adhesive bonding system. The retentive pattern produced on the glass-ceramic surface is particularly advantageous in this respect.

It was possible to develop a leucite apatite glass-ceramic derived from the SiO_2-Al_2O_3-Na_2O-K_2O-CaO-P_2O_5-F system by combining two different mechanisms, i.e. controlled surface nucleation and controlled bulk nucleation. IPS d.SIGN is amongst these. The glass-ceramic was prepared according to the classic method of glass-ceramic formation: melting, casting to prepare a glass frit, controlled nucleation and crystallization. A two-fold reaction mechanism leads to the precipitation of fluoroapatite, $Ca_5(PO_4)_3F$ and leucite, $KAlSi_2O_6$. SEM pictures show the two-phase crystal content of apatite and leucite in this type of glass-ceramic. Fluoroapatite phase takes the form of needle-shaped crystals whereas the oval areas are the leucite crystals. The clinical application of this glass-ceramic has been proven to be suitable for clinical application as veneering material on metal frameworks for single units as well as for large dental bridges involving more than three units.

The first glass-ceramic to be developed was by Stookey et al. which contained Lithium disilicate. Further research into this field allowed for IPS Empress 2 to be developed. This glass-ceramic was developed in order to extend the range of indications of glass-ceramics from inlay and crowns to three-unit bridges, by offering high strength, high fracture toughness and at the same time, a high degree of translucency. Both the flexural strength and fracture toughness of lithium disilicate glass-ceramics are almost three times of those of leucite glass-ceramics. Lithium disilicate glass-ceramic ingots are utilizied to produce the crown or bridge framework in combination with the viscous flow process. To further improve the aesthetic properties, i.e. translucency and shade match, and to optimally adjust the wear behavior to that of the natural tooth, the lithium disilicate glass ceramic is veneered with an appetite-containing glass-ceramic using a sintering process.

In order to meet the demanding requirements of CAD/CAM applications, a lithium metasilicate glassceramic, IPS e.max was developed. This material, which is supplied in a typically blue color, is adjusted by thermal treatment in order to demonstrate a characteristic tooth color.

The range of IPS e.max products also encompasses various apatite-containing glass ceramics that are suitable for both layering material on lithium disilicate glass-ceramic and veneering material on ZrO_2 sintered ceramic. The apatite crystal phase of the $Ca_5(PO_4)_3F$ type acts as a component that adjusts the optical properties of the restoration to natural tooth. For this reason, the crystallites are of nanoscale dimension.

ZrO_2 containing glass-ceramics was the first glass-ceramic developed to be fused to high strength ZrO_2 ceramic dental posts. The glass-ceramic contains $Li_{12}ZrSi_6O_{15}$ crystals as the main phase; however different types of crystals are also precipitated in the glassy matrix. ZrO_2 has become very interesting not only in the field of medicine but also in dental applications. High-strength and high toughness dental posts, crowns and bridges can be prepared from this material.

In order for a dental restorative material to be of clinical success, their most important properties include; high strength, high toughness, abrasion behavior comparable to natural teeth, translucency, color, durability) and the processing technologies (moulding, machining, sintering). Furthermore, the material should have good marginal fit with the tooth, biocompatibility, good mechanical properties and low porosity. In addition to the aforementioned properties, the recent requirement for dental restorative materials is for its appearance to be similar to that of a natural tooth.

Glass-ceramics have been researched immensely in order to fulfil high standards of function and aesthetics from an early stage. The trend for metal free dental restorations began from the 1970's whereby metal free feldspathic ceramics were reinforced with additional components. Since then, increasing the strength of these materials progressed rapidly by controlling the nucleation and crystallisation of glasses. These developments have now led to the introduction of a trend which is focused on achieving exceptional aesthetic results with glass ceramics as metal free dental restorations.

Although glass-ceramics exhibit the desired properties for dental restoration, their main drawback is that they are brittle which the main cause of failure is. This is due to either fabrication defects; which are created during production of the glass-ceramic or secondly, surface cracks; which are due to machining or grinding. Therefore when processing the glass-ceramic, care needs to be taken in addition to choosing the suitable method for production for specific compositions of the glass-ceramic in order to improve their mechanical properties.

Apart from the use of glass-ceramics for dental restorations, they can also be applied as bioactive materials for the replacement of hard tissue. Bone is a complex living tissue which has an elegant structure at a range of different hierarchical scales. It is basically a composite comprising collagen, calcium phosphate (being in the form of crystallised hydroxyapatite, HA or amorphous calcium phosphate, ACP) and water. Additionally, other organic materials, such as proteins, polysaccharides, and lipids are also present in small quantities. Because bone is susceptible to fracture; there has always been a need, since the earliest time, for the repair of damaged hard tissue.

Many years of research has attempted to use biomaterials to replace hard tissue, ranging from using bioinert materials, to bioactive materials such as 'Bioglass' to 'Apatite-wollastonite (A-W) glass-ceramics and to calcium phosphate materials. Calcium phosphate based materials have received a great deal of attention in this field due to their similarity with the mineral phase of bone.

Calcium Phosphate Glasses

The application of calcium phosphate material as a bone substitute began by Albee, who reported that a tricalcium phosphate compound used in a bony defect promoted osteogenesis. Many years later, Levitt et al. and Monroe et al. were the first to suggest the use of calcium phosphate ceramics for dental and medical implant materials. Subsequently in 1971, Hench et al developed a calcium phosphate containing glass-ceramic, called Bioglass and demonstrated that it chemically bonded with the host bone through a calcium phosphate rich layer. Furthermore the advantageous properties of calcium phosphate ceramics arose when Nery et al. used a calcium phosphate ceramic for implants in surgically produced infrabony defects in dogs. This demonstrated that the calcium phosphate ceramic was nontoxic, biocompatible, and caused no significant haematological changes in the calcium and phosphorus levels. Since then, a great deal of research into calcium phosphate glass-ceramics has been conducted as potentially bone substitutes in dentistry.

Calcium phosphate based ceramics can be characterized accordingly:

1. Hydroxyapatite (HA, $Ca_5(PO_4)_3OH$),
2. β-tricalcium phosphate (β-TCP, β-$Ca_3(PO_4)_2$),
3. Biphasic calcium phosphates, BCP; mixture of HA and β-TCP,
4. β-calcium pyrophosphate (β-CPP, β-$Ca_2P_2O_7$),
5. Fluorapatite (FAP, $Ca_5(PO_4)_3F$).

Calcium phosphate based ceramics and their properties have been characterized according to the proportion of calcium to phosphorus ions in the structure. One of the most widely used synthetic calcium phosphate ceramics is hydroxyapatite, $Ca_5(PO_4)_3OH$, HA and this is due to its chemical similarities to the inorganic component of hard tissues. HA, has a Ca:P molar ratio of 1.67. It has higher stability in aqueous media than other calcium phosphate ceramics.

Tricalcium phosphate (TCP) is a biodegradable bioceramic with the chemical formula, $Ca_3(PO_4)_2$. TCP dissolves in physiological media and can be replaced by bone during implantation. TCP has four polymorphs, the most common ones being α and β-forms, of which β-TCP has received a lot of attention in the field of bone substitutes. Slight imbalances in the ratio of Ca:P can lead to the appearance of extraneous phases. If the Ca:P ratio is lower than 1.67, then alpha- or beta tricalcium phosphate may be present after processing. If the Ca:P is higher than 1.67, calcium oxide (CaO) may be present along with the HA phase. These extraneous phases may adversely affect the biological response to the implant in-vivo. A TCP with a Ca:P ratio of 1.5 is more rapidly resorbed than HA. Hence, β-TCP has been involved in recent developments aimed to improving its biological efficiency and its mechanical properties in order for it to be successful as bone substitutes.

Mixtures of HA and TCP, known as biphasic calcium phosphate (BCP), have also been investigated as bone substitutes and the higher the TCP content in BCP, the higher the dissolution rate. The crystal structure of HA can accommodate substitutions by various other ions for the Ca^{2+}, $PO4_3^-$ and OH^- groups. The ionic substitutions can affect the lattice parameters, crystal morphology, crystallinity, solubility and thermal stability of HA. Anionic substitutions can either occur in the phosphate- or hydroxyl positions. Fluorapatite and chlorapatite are common examples of anionically substituted HA. They display a similar structure to HA, but the F^- and Cl^- ions substitute for OH^-. A lot of research has gone into carbonate substituted HA and it has shown to have increased bioactivity compared to pure HA, which is attributed to the greater solubility of the carbonated substituted HA. Thus, recent work has been in progress in order to optimise the production and sintering behavior of carbonated substituted HA in order for use in biomedical applications.

Materials which are bioactive i.e. the ability to bond to living tissue and enhance bone formation, have the following characteristic compositional features: (i) SiO_2 contents smaller than 60 mol%, (ii) high Na_2O and CaO content, and (iii) high CaO:P2O5 ratio. Although silica based bioactive materials have shown great clinical success in many dental and orthopaedic applications, its insoluble properties has resulted in it as a potential for a long term device and the long term reaction to silica, both locally and systematically is still unknown. Therefore, silica free, calcium phosphate glasses have attracted much interest due to their chemical and physical properties. They offer a more controlled rate of dissolution compared to silica containing glasses, they are simple, easy to produce, biodegradable, biocompatible, bioresorbable due to their ability to completely dissolve in an aqueous environment and have excellent bioactivity, osteoconductivity as well as not causing an inflammatory response. Due to their properties, especially due to it being bioresorbable, calcium phosphate glasses have been under investigation for several applications in the dental field, particular as implants. However only certain calcium phosphate compounds are suitable for implantation in the body, compounds with a Ca:P ratio less than 1 are not suitable for biological implantation due to their high solubility.

The structural unit of phosphate glasses is a PO_4 tetrahedron. The basic phosphate tetrahedra form long chains and rings that create the three-dimensional vitreous network. All oxygens in the glass structure are bridging oxygens (BO), and the non-bridging oxygens (NBO) can be formed by including other species such as CaO and Na_2O or MgO. Do to the effects of Ca^{2+}, Na^{2+} and Mg^{2+} in the glass structure; they are defined as glass network modifiers, which form the glassy state and are called 'invert glasses.' Hence the structure of phosphate glasses can be described using the Qn terminology, where n represents the number of bridging oxygen's that a PO_4 tetrahedron has in a P_2O_5 glass, every tetrahedron can bond at three corners producing layers of oxygen polyhedra which are connected together with Van der Waals bonds. When the PO_4 tetrahedron bonds with three bridging oxygens, giving the Q3 species, it is referred to as an ultraphosphate glass, which usually consists of a 2D network. When it bonds to two bridging oxygen's, usually in a 3D-network it gives the Q2 species, it is referred to as metaphosphate glass. Further addition gives Q1 species, also called pyrophosphate glass, which bonds only to one bridging oxygen. Finally, the Qo species do not bond to any bridging oxygen and hence is known as an orthophosphate glass.

A large number of calcium phosphate glass compositions have been studied in order to exhibit suitable properties for use in biomedical applications until now, and they can be categorised into four groups:

1. Calcium phosphate glasses containing Potassium.
2. Calcium phosphate glasses containing Magnesium.
3. Calcium phosphate glasses containing Sodium and Titania.
4. Calcium phosphate glasses containing Fluorine and Titania.

Calcium Phosphate Glasses Containing Potassium

Dias et al. conducted a study and prepared bioresorbable calcium phosphate glass-ceramics between the metaphosphate and pyrophosphate region based on the composition $45CaO$-$45P_2O_5$-$5K_2O$-$5MgO$ (Ca:P = 0.5). XRD results showed that addition of nucleating agents, K_2O and MgO forms bioactive: β-CPP and biodegradable phases: $KCa(PO_3)_3$, $Ca_4P_6O_{19}$ as well as β-$Ca(PO_3)_2$ which is considered to be non-toxic. DTA results showed two crystallisation peaks, Tp at 627 °C and 739 °C and two melting temperatures, Tm at 773 °C and 896 °C which was thought to be due to the partial melting of the crystalline phases or residual glass matrix. The glass transition temperature, Tg was observed at 534 °C. FTIR results showed functional groups corresponding to metaphosphate and pyrophosphate, $(PO_3)^-$ and $(P_2O_7)^{4-}$. These results are in accordance with functional groups of the crystalline phases identified by XRD: β-CPP, $KCa(PO_3)_3$, $Ca_4P_6O_{19}$ and β-$Ca(PO_3)_2$. Results from degradation studies of these glass-ceramics confirmed that by controlling the overall composition of the O:P in the glass, glass ceramics with high degree of degradability can be obtained. The level of chemical degradation observed for these materials is well-above that reported in literature for bioactive ceramics that are clinically used, namely HA and TCP. It was therefore concluded that the incorporation of K_2O in glass ceramics increases the solubility and also these calcium phosphate glass ceramics makes them potentially clinically helpful for promoting the regeneration of soft as well as hard connective tissue by allowing the degradability to be controlled.

A study by Knowles et al. investigated the solubility and the effect of K_2O in the glass-system based on the general composition: K_2O-Na_2O-CaO-P_2O_5. The exchange of a mono or divalent ion with another of a similar charge was therefore investigated. The P_2O_5 and CaO content were fixed, at 45 mol% and the CaO content at 20, 24 or 28 mol% and the ratio of K_2O to Na_2O was varied from 0 to 25 mol%. Results showed, firstly an increase in CaO content caused the solubility to decrease, as expected and confirmed from previous studies. Secondly, for all CaO contents there was an increase in solubility, when K_2O content was increased. In a recent study by Marikani et al. based on the same general composition, they demonstrated that the addition of K_2O caused a decrease in both density (from 2.635 g cm^{-3} to 2.715 g cm^{-3} and microhardness measurements (from 257 to 335 HV) and hence weakens the structure. These findings are attributed to the replacement of lighter cation (Na_2O) by a heavier one (K_2O). The ionic radius of potassium is larger than the ionic radius of sodium so, the addition of K_2O has a larger disrupting effect on the structure and hence weakens the glass-network. The decrease of melting point with the addition of K_2O content indicates that K_2O increases network disruption by producing non-bridging oxygens. And the low value of Tg indicates that the glass samples are thermally unstable. Additionally, the elastic modulus, decreases when the concentration of K_2O is increased, which implies the weakening of the overall bonding strength, as more cross linking is degraded. The increase of the internal friction and the decrease of the thermal expansion coefficient with the addition of the K_2O content are due to the formation of non-bridging oxygen ions. The SEM micrographs of the glass samples recorded before immersion in SBF indicates the amorphous nature of the materials and when glasses were immersed in SBF solutions for 10 days, the glass-samples showed bioactivity.

Although the addition of K_2O to the ternary Na_2O-CaO-P_2O_5 based system offers greater flexibility in terms of tailoring the solubility to suit potential biomedical applications, only little research has been conducted in using K_2O in calcium phosphate glasses, probably because it has shown to increase network disruption which was confirmed by decrease in Tm, addition of K_2O causes a decrease in density and microhardness measurements, it weakens overall bonding strength confirmed by a decrease in the elastic modulus, causing it to be less rigid as well as producing thermally unstable glasses which was confirmed by the low values of Tg. These mechanical properties are not desirable in the long run and due to it being less rigid, it would not withstand stress in biomedical applications and consequently result in failure.

Calcium Phosphate Glasses Containing Magnesium

Research into calcium phosphate glasses which produce biocompatible and bioactive phases has generated a lot of interest. An attempt to induce β-TCP was undertaken by Zhang et al. on calcium phosphate glass-ceramics in the pyrophosphate region based on the composition $50CaO-40P_2O_5$-$7TiO_2$-1.5MgO-1.5Na_2O (Ca:P molar ratio = 0.625). XRD results showed that the β-TCP phase was not detected and the main crystalline phase precipitated was β-CPP with smaller amounts of soluble Calcium titanophosphate, $CaTi_4(PO_4)6$ CTP, and Sodium titanophosphate, $NaTi_2(PO_4)_3$. Kasuga et al. reported a similar occurrence in the structure of glass-ceramics which contained TiO_2 (wt 3%). SEM observations demonstrated light areas which were confirmed by EDS analysis to be β-CPP, grey areas was thought to correspond to Na- containing phases and dark areas were composed of lower CaO contents compared to the other two areas and contained MgO and Na_2O. These results were identical to Kasuga et al's study. The undetectable β-TCP phase was possibly due to the low content of MgO and TiO_2 added and the low Ca:P ratio of the glass. Although bioactive and biosoluble phases were precipitated in the glass-ceramic, no continuous apatite layer was formed even after 8 weeks of immersion in SBF solution.

A study by Brauer et al. observed the solubility of several phosphate glasses in the system P_2O_5-CaO-MgO-Na_2O-TiO_2. The glass compositions ranged from ultraphosphate glasses (with phosphate contents over 50 mol%) to polyphosphate glasses (containing 50 mol% P_2O_5 or less which are formed by phosphate chains or rings possessing different chain lengths) to invert glasses (pyrophosphate glasses-P_2O_5 concentrations of around 34 mol%). Results showed that the phosphate glasses showed a uniform dissolution. No selective alkali leaching, which is known from silica based glasses, was observed. Also that the solubility of the glasses strongly depend on the glass-composition. The higher the phosphate content resulted in an increase in solubility. According to Vogel et al., this is due to the polymerisation of the phosphate chains and the Q1 end units being more susceptible to hydration and subsequent hydrolysis than Q2 middle groups. Also it was observed that the higher the concentration of Na_2O resulted in an increase in solubility too due to the effect Na^+ has on the glass structure. Addition of titanium oxide resulted in a decrease in both the solubility and the tendency of the glasses to crystallise by forming cross links between phosphate groups and titanium ions. Invert glasses showed a considerably smaller solubility than polyphosphate glasses and offer an alternative to polyphosphate glasses, since they are more stable to moisture attack. However, decreasing the P_2O_5 content makes glasses not only more stable to hydrolysis but also restricts the glass forming area. Hence, glasses in the pyrophosphate region show a larger tendency to crystallize than polyphosphate glasses. However invert glasses in the system P_2O_5-CaO-MgO-Na_2O showed that properties such as solubility and crystallization tendency can be controlled by adding small amounts of metal oxides. Results of solubility experiments showed that the glass system investigated enabled adjustment of solubility with only minor chemical changes. This ability to control the solubility is very promising for medical application where the coordination of implant degradation and bone formation are a key issue.

Ceramic Cements

Ceramic cements help isolate, insulate and protect components in a variety of applications. Specialty ceramic cements are directly relevant to our lives, but we rarely notice their presence. The important physical characteristics of these cements include the capacity to deal with temperature changes varying from ambient temperature up to 3000 °F (1650 °C) and higher, depending on the specific case.

Ceramic cements help to keep the different components that form a particular par.

Important physical characteristics of ceramic cements include the capacity to deal with temperature changes varying from ambient temperature up to 3000 °F (1650 °C)and higher.

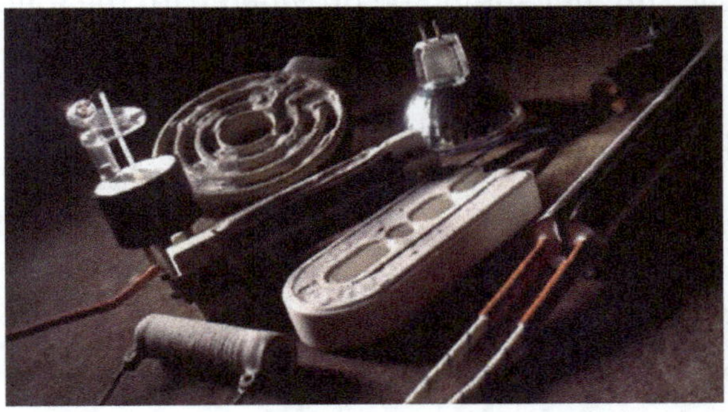

Each material has a different coefficient of thermal expansion. Selecting the cement with the proper coefficient of thermal expansion (i.e., one that is similar to that of the other material in the application) is important to avoid cracks and failures.

Properties

Some ceramic cement are insulators; they are selected not only for their temperature resistance, but also for their ability to insulate and protect parts from damage due to extreme high temperatures, like in thermocouples used to check the quality of molten iron and other molten ores. Some ceramic cements are also chemical resistant, making them appropriate for service in acidic or alkaline environments. In other words, they are not only able to withstand the temperature changes and aggressive environments, but they can also be used as insulators or in contact with dissimilar materials. In addition, ceramic cements are able to isolate parts of an application from the surroundings. The combination of all of these properties helps the pieces and parts to do their jobs.

Nearly all ceramic cements are made of inorganic an material, which presents a small problem: porosity. Some ceramic cement are never going to be able to protect the part in full from water and humidity penetration. This is characteristic of inorganic ceramic cements. In many cases, the addition of a primer helps to delay or prevent this kind of penetration.

Some of these cements are made by mixing the cement with water, which is required for the material to cure. In the curing process, it is crucial to eliminate all the excess water (also known as water of convenience) used to hydrate the cement and initiate the curing process. After the cement has cured and reached its peak, that water of convenience needs to be removed before the part is put in service to avoid cracking, steaming or even short-circuiting the part (if it is transmitting electricity).

Some ceramic cement need to be fast curing to comply with manufacturers' demands to be able to assemble and pack the parts in order to send to consumers as soon as possible. Others need to be oven baked to cure properly, letting all the water used in the mixing of the cement reach a complete cure to avoid problems when the parts are put into service.

Selection

When selecting the proper cement, keep in mind that some ceramic cement might have a bad reaction when in contact with certain metals and other parts. To select the proper ceramic cement for your application, the ceramic supplier should be provided with the following details:

- What the application is;
- Degree of electrical resistance required;
- Temperature range the cement will withstand;
- Thermal shock resistance required;
- How fast the temperature goes from one extreme to the other;
- Frequency of those temperature changes;
- Desired thermal conductivity;
- Allowed degree of thermal expansion;
- What materials will be in contact with the cement;
- Whether volume stability is required;
- Whether moisture absorption will be a problem;
- Dispensing methods;
- Required pot life;
- Whether a predetermined set time exists;
- Whether a force cure is needed.

Ceramic cements can be applied in many different ways: potting, casting, sealing, encapsulating, bonding and coating. They help to keep the different components that form a particular part isolated, insulated and protected. They help to keep all the pieces together, bonding metals to ceramics and to glass, to avoid the de-bonding of the components.

Uses of Ceramic Cements

In Garage

Every breaker in your garage's electric panel is assembled using specialty ceramic cement able to keep the calibration of the breaker in the right position. All of the international assemblers of those products use this kind of cement.

In Kitchen

In your kitchen, gas ranges with electric starters are all made with special ceramic cement that can take the heat generated to ignite the gas for you to fry an egg or boil water. By pushing a button or turning a knob on your gas range, a spark jumps from the edge of the burner into the path from where the gas is coming to ignite the gas and provide flame for the burner.

When you choose to toast a bagel for breakfast, all the resistors located in the toaster are assembled using specialty ceramic cement able to withstand the high temperature required to toast your bagel to perfection.

Like many families, you probably have a set of silverware that was passed down through the generations. Many people believe these antique pieces are made entirely of silver, but the truth is that all of them are assembled using a special ceramic cement to give weight and balance to the piece of silverware and make it affordable to everyone.

In Car

Every light bulb on your dashboard is assembled with specialty cement able to take the cycling of the temperature from the moment you start your car until you turn it off.

At Mall

Did you ever see store window displays that are illuminated by tiny little light bulbs (halogen lamps)? Those lamps are assembled using ceramic cement that is able to handle not only the high temperature, but also the electrical conductivity necessary to keep the lamps lit.

Structural Ceramics

Structural ceramics are usually formed by extrusion, sometimes followed by plastic pressing, whereas plastic pressing is typically used in forming roof tiles and vases, and to a lesser extent in forming bricks and special pieces.

Structural ceramics are generally brittle, which seriously affects the reliability of ceramic components. Four methods are typically used to resolve this weakness: (a) fiber reinforcement or microstructural control to toughen the ceramic, (b) high-level non-destructive inspection and repair of the dangerous flaws, (c) proof testing to select components with high reliability, and (d) using materials with self-crack-healing ability.

Cyclic fatigue performance of structural ceramics determines the life of components in many cases, including gas turbine rotors or blades and diesel engine valves. Therefore, characterizing and properly understanding ceramic cyclic fatigue is crucial for confident use of components subjected to cyclic loading. Many reports explain cyclic fatigue of ceramics using short or natural flaws.

Several mechanisms for cyclic fatigue of ceramics have been proposed. Most of them are related strongly to toughening mechanisms of ceramics at room temperature. They include degradation of transformation toughening zone at the crack tip (such as in some zirconia); damage to bridging reinforcements (grains, whiskers, platelets, and brittle or ductile particles) over the bridged interface; cyclic loading-induced accumulation of microcracks around the crack tip; crack tip blunting/resharpening; and relaxation of any mechanically beneficial residual stress due to microcracking, as shown in figure. It should be noted that these degradation mechanisms are not active in static or dynamic fatigue.

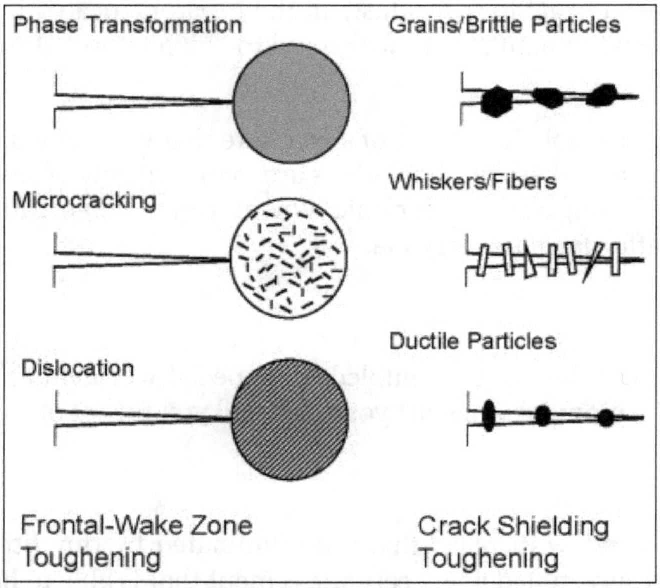

Schematic classification of primary microstructural toughening mechanisms, which are strongly involved with property degradation during cyclic loading.

In some cases, at elevated temperatures where softening of the grain boundary glassy phase can occur, other mechanisms become active. These include oxidation-induced deformation and visco-elastic effects.

Lead zirconate titanate (PZT) ceramics have potential for increased use in advanced industrial applications such as piezoelectric actuators and ultrasonic vibrators because of their quick response, compactness, and good power efficiency. When they are subjected to a powerful alternating electric field, cyclic stresses are introduced into these piezoelectric ceramics which can lead to fracture and microcracking. Moreover, degradation of piezoelectric properties is a serious practical concern, as this is a major cause of reduced device lifetime. Therefore, in recent years, many studies have been devoted for clarifying cyclic fatigue behavior and mechanical reliability of this kind of material.

Ceramic materials have been also extensively applied in prosthodontics in recent years, since all-ceramic crowns are superior to traditional porcelain-fused-to-metal crowns in esthetics, wear resistance, and chemical inertness. In such applications, the materials are susceptible to cyclic fatigue that substantially degrades their mechanical reliability and structural durability. Therefore, considerable efforts have been made to investigate cyclic fatigue behavior of various ceramics that can be used for these applications.

Advanced Ceramics

Advanced ceramics such as alumina, aluminum nitride, zirconia, silicon carbide, silicon nitride and titania-based materials, each with their own specific characteristics, offer a high-performance, economic alternative to conventional materials such as glass, metals and plastics.

The demand posed by new and changing applications is to improve operation at a reduced cost. New materials are being continuously engineered and adapted in order to address the needs of specific and often unique applications.

Joining ceramics to metals creates its own engineering challenges that require specialist expertise. Morgan Advanced Materials is a global leader in the metallisation and joining of ceramics. Its application engineers have worked with customers all over the world to provide high-integrity solutions for components of all sizes, shapes and specifications.

Desired Properties for the Application

Physical properties such as hardness, strength, wear resistance; corrosion resistance and thermal stability are considered while choosing a material. Each of these can be optimized depending on the choice of material.

Materials

Designers need to consider the various ceramic materials available to them. Several ceramic materials are favored and have a proven track record for their mechanical, electrical, thermal and chemical properties. Examples are detailed below:

1. Alumina: Alumina is a versatile material that offers a combination of good mechanical and electrical properties. It is suited for a variety of applications, which include X-ray tubes, electron tubes, laser devices, aerospace devices, high vacuum applications, flow meters, pressure sensors and wear components.

It has good stiffness and strength, good resistance to wear and high hardness. Alumina is offered in many grades ranging from 60% to 99.9% with additives designed to enhance properties such as wear resistance or dielectric strength. It can be formed using several ceramic processing methods and can be processed machined or net-shaped to produce a variety of sizes and shapes. Furthermore it can be readily joined to other ceramics or metals using specially developed metallising and brazing techniques.

2. Aluminum Nitride (AlN): Aluminum nitride (AlN) exhibits very good thermal conductivity. Other properties include excellent thermal shock resistance and corrosion resistance. Based on these properties AlN is used in power electronics, aeronautical systems, railways, opto-electronics, semiconductor processing, microwave and military applications. Typical applications include heaters, windows, IC-packages and heat sinks.

3. Zirconia: Zirconia offers corrosion and chemical resistance at high temperatures up to 2400 °C – well above the melting point of Alumina. Magnesia Partially Stabilised Zirconia (Mg-PSZ) and Yttria Tetragonal Poly-crystal Zirconia (Y-TZP) are suited to engineering or structural applications where exceptional mechanical strength and properties such as hardness, wear and corrosion resistance are required.

The high temperature capability of zirconia products has resulted in the development of Fully Stabilised Zirconia (FSZ) grades for crucibles, nozzles and other components for molten metal handling applications.

4. Silicon Nitride: Silicon nitride has very good high-temperature strength, creep and oxidation resistance, while its low thermal expansion coefficient provides good thermal shock resistance when compared with most other ceramic materials.

It has high fracture toughness, hardness, chemical and wear resistance, and is manufactured in three main product types: reaction bonded silicon nitride (RBSN), hot pressed silicon nitride (HPSN) and sintered silicon nitride (SSN). Typical applications include: bearing ball and roller elements, cutting tools, valves, turbocharger rotors for engines, glow plugs, non-ferrous molten metal handling, thermocouple sheaths, welding jigs and fixtures and welding nozzles.

5. Silicon Carbide: Silicon carbide is highly wear-resistant with good mechanical properties including high temperature strength and thermal resistance of up to 1650 °C. It has low density, high hardness and wears resistance and excellent chemical resistance. The applications of SiC are fixed and moving turbine components, seals, bearings, ball valve parts and semiconductor wafer processing equipment. A significant area of use is in specialist thermal processing applications, including beams and profiled supports, rollers, tubes, batts and plate setters, as well as thermocouple protective sheaths.

Shape of the Components

Once the material has been decided, the shape is the next consideration. There are certain shapes that will cause weaknesses in the component. When working with ceramic, simple shapes are consistently seen to provide the strongest result.

Metallising

In some applications there is a requirement to join ceramic to metal to create the finished part. Various methods are available including mechanical fasteners, friction welding and adhesive bonding, but by far the most widely used and effective method for creating leak-tight, robust joint between ceramic and metal is brazing. This begins with the chemical bonding of a metallisation layer on the ceramic to create a wettable surface upon which the braze alloy will flow between the two components during the brazing process.

Morgan Advanced Materials manufactures WESGO braze alloys and supplies high-purity, low vapour pressure alloys, including precious metal filler materials, non-precious alloy filler materials and active braze alloys. Precious brazing filler metals are derived from gold, silver, platinum and palladium—based materials and exceed the most stringent requirements imposed by the power tube, aerospace, and semiconductor, medical, electronic and vacuum industries which they serve. Non-precious alloy filler materials are ideal for applications including tooling for mining and heavy industry equipment. They are suitable for brazing applications between 500 °C and 1200 °C.

Co-fired Assemblies

A metal feed through can be produced for specific applications, such as flow meters, by placing a wire in the ceramic in the pre-sintered (green) stage. As the ceramic shrinks during the sintering process, it compresses on the metal and forms a gas-tight seal.

Coating and Glazing

The roughness of a final product is based on the grain size. If the grain size is large, then the product will have a rough finish which can lead to cavities being formed after grinding. In order to achieve an excellent surface finish, parts can be glazed.

Alumina Ceramics

Alumina or Aluminium Oxide, Al_2O_3 is a major engineering material. It offers a combination of good mechanical properties and electrical properties leading to a wide range of applications. Alumina can be produced in a range of purities with additives designed to enhance properties. A wide variety of ceramic processing methods can be applied including machining or net shape forming to produce a wide variety of sizes and shapes of component. In addition, it can be readily joined to metals or other ceramics using metallising and brazing techniques.

Alumina based ceramics are by far the largest range of advanced ceramics made by Morgan Technical Ceramics. Due to the important combination of properties, we have thoroughly researched the behavior and characteristics of our Alumina products to give you the best possible component.

Typical Alumina characteristics include:

- Good strength and stiffness,
- Good hardness and wear resistance,
- Good corrosion resistance,
- Good thermal stability,
- Excellent dielectric properties (from DC to GHz frequencies),

- Low dielectric constant,
- Low loss tangent.

Typical Alumina applications include:

- Seal rings,
- Medical prostheses,
- Laser tubes,
- Electronic substrates,
- Ballistic armour,
- Thermocouple tubes,
- Electrical insulators,
- Grinding media,
- Thread guides,
- Wear components.

Electroceramics

Electroceramics are advanced ceramic materials that are employed in a wide variety of electric, optical, and magnetic applications. In contrast to traditional ceramic products such as brick and tile, which have been produced in various forms for thousands of years, electroceramics are a relatively recent phenomenon, having been developed largely since World War II. During their brief history, however, they have had a profound impact on the so-called electronics revolution and on the quality of life in developed nations. Electroceramics that have low dielectric constants (*i.e.*, low electric resistivity) are made into substrates for integrated circuits, while electroceramics with high dielectric constants are used in capacitors. Other electroceramic materials exhibit piezoelectricity (the development of strain under an applied field, or vice versa) and are employed in transducers for microphones and other products, while some possess good magnetic properties and are suitable for transformer cores or permanent magnets. Some electroceramics exhibit optical phenomena, such as luminescence (useful in fluorescent lighting) and lasing (exploited in lasers), and still others exhibit changes in optical properties with the application of electric fields and are therefore used extensively as modulators, demodulators, and switches in optical communications.

All the applications listed above require electric insulation, a property that has long been associated with ceramics. On the other hand, many ceramics are suitable for doping by aliovalent materials (that is, materials with other charge states than the ions of the host crystal). Doping can lead to electrically conductive ceramics, which appear in products such as oxygen sensors in automobiles, heating elements in toaster ovens, and transparent oxide films in liquid crystal displays.

In addition, ceramics have been developed that are superconducting; that is, they lose all electric resistivity at cryogenic temperatures. Because their critical temperatures (T_c's; the temperatures at which the transition occurs from resistivity to superconductivity) are much higher than those of conventional metallic superconductors, these ceramic materials are referred to as high-T_c superconductors.

Most electroceramics are truly high-tech materials, insofar as they are made into high value-added items. Starting materials of high purity are employed, often in clean-room processing facilities. Because grain size and grain size distribution can be the deciding factors in the quality of the electroceramic being produced, strict attention is given to the steps of powder processing, consolidation, and firing in order to achieve the desired microstructure. The structure and chemistry of grain boundaries (the areas where two adjacent grains meet) must often be strictly controlled. For example, the segregation of impurities at grain boundaries can have adverse effects on ceramic conductors and superconductors; on the other hand, some ceramic capacitors and visitors depend upon such grain boundary barriers for their operation.

Piezoelectric Ceramics

A piezoelectric ceramic is a smart material that converts a mechanical effect (such as pressure, movement, or vibration) into an electrical signal and vice versa. Due to the electromechanical effect, piezoelectric ceramics are used in a wide range of applications such as motion sensors, watches, ultrasonic power transducers, lithotripters, ultrasonic cleaning, ultrasonic welding, active vibration dampeners, high-frequency loudspeakers, actuators for atomic force microscopes, and many others. Lead zirconate titanate (PZT), barium titanate (BT), and strontium titanate (ST) are the most widely used piezoelectric ceramic materials.

The piezoelectric nature of ceramic materials can be evaluated by measuring parameters such as the piezoelectric charge coefficients (d constant), the piezoelectric voltage coefficients (g), and the piezoelectric coupling factors (k). Due to the anisotropic nature of piezoelectric ceramic materials, constants d and g of the ceramic materials relate to both in the direction and the directions perpendicular to the applied electric field or mechanical force. The d constant of ceramic materials is the ratio between the mechanical strain developed per unit of electric field or polarization experienced per unit of mechanical stress, applied to a piezoelectric material. Whereas, the g constant of ceramic materials can be defined as the mechanical strain developed per unit of electric displacement, or electric field generated per unit of mechanical stress, applied to a piezoelectric material. High d is desirable in ceramic materials used as actuators, such as in an AFM, where well-controlled motion and vibration are required during surface scanning. The g is related to the d by the following equation:

$$d = \varepsilon g$$

where, ε is the relative permittivity or dielectric constant of piezoelectric ceramic materials and is defined as the dielectric displacement per unit electric field. The piezoelectric charge coefficients (d) can be measured by means of quasistatic and interferometric methods. In the quasistatic technique, a static force is applied to a specific area of the piezoelectric ceramic materials. The d is the ratio between the charge generated (Q) in ceramic materials to the static force (F) applied to a defined area of the ceramic materials. However, the application of the quasistatic method is limited in many cases as the measurement does not consider frequency dependencies and neglects the charge/polarization effect from the space charges. In the interferometric technique, the change of the piezoelectric ceramic materials is detected as a function of applied AC voltage between two ceramic materials. However, this technique usually requires a reflecting mirror like surface and therefore can only be used for ceramic materials that exhibit mirror like surface. A material's piezoelectricity can also be assessed with the dynamic, resonance, and acoustical methods to overcome the limitation described earlier. For example, in the dynamic method, the charge signal of a piezoelectricity effect is periodic at a controlled frequency, when a periodic force is applied. In this technique, the charge signal due to the piezoelectricity can be easily separated from any other charge signal contributions.

The high piezoelectric voltage coefficients are required in materials intended for use as sensors and to generate voltage in response to the mechanical energy. The overall 'strength of the electromechanical effect' of piezoelectric ceramic materials can be expressed by the piezoelectric coupling factor. The piezoelectric coupling factors (k) can be defined by the following equation:

$$k\sqrt{\frac{\text{Electrcal Energy Output}}{\text{Total Mechanical Energy Input}}}=\sqrt{\frac{\text{Mechanical Energy Output}}{\text{Total Electrical Energy Input}}}$$

Piezoelectric ceramics, such as lead zirconate titanate (PZT) and barium titanate ($BaTiO_3$), have been widely used as sensors and actuators. Recently, they have also gained popularity for use in energy harvesting devices. Their inherently high electromechanical coupling properties are appealing for actuator applications, where relatively small electric field inputs are sufficient to actuate the ferroelectric ceramics. However, the brittle nature of ceramics limits their applications to only small deformations. In electromechanical devices, several characteristics may be required, such as light weight, high electromechanical coupling constants, low thermal expansion and conductivity, mechanical flexibility and compliance, etc. For this purpose, electroactive composites with several different constituents have been considered. Newnham et al. discussed the key features in achieving desired properties in developing active composites, which are tailored to the arrangement of the constituents (connectivity). The most common and practical types are composites with active piezoceramic inclusions of particles or long fiber shapes dispersed in continuous soft matrix, for example, polymers. These composites are referred to as 0–3 and 1–3 composites, respectively.

Piezoelectric composites are generally fabricated by embedding non-polarized piezoelectric ceramics into a passive soft matrix, such as polymers and metals. By adjusting spatial concentration and geometry of the piezoelectric inclusions, one can fine tune material properties of the composites to specific device requirements. The composites having piezoelectric particles and polymer matrix are then polarized, which can be quite challenging because the passive polymeric matrix has very low dielectric constants and typically the ceramics inclusions do not form a continuous connection across the composites. Cui et al. discussed the fact that polarization of the composites with piezoelectric particles can be improved by having a composite with high particle percolations, and reducing the ratios of the dielectric constants of the ceramics and polymers. Several experimental studies have shown that piezoelectric composites exhibit more desirable performance than monolithic piezoelectric showed that piezocomposites with unidirectional PZT fibers have high electromechanical coupling constants while they are compliant.

The overall electromechanical properties and behaviors of piezoelectric composites are strongly influenced by the properties and behaviors of the constituents, that is, inclusions and matrix, microstructural morphologies of the composites such as shape, size, and arrangement of the inclusions, and the volume or weight contents of the constituents. The response of the piezoelectric ceramics also depends on the loading conditions that the piezoelectric ceramics are subjected to. When polarized piezoelectric ceramics are subjected to relatively small electric field inputs, a linear response in stress/strain and electric flux is observed. However, relatively large electric field inputs, even below the coercive electric field limit, lead to nonlinear electromechanical responses; see, for example. In such situations, the linear piezoelectricity that was standardized by the IEEE is no longer applicable to describe the electromechanical behavior of the piezoelectric ceramics. Experimental evidence shows that the electrical and mechanical responses of the polarized piezoelectric ceramics are time dependent. When subjected to cyclic electric fields with amplitude less than the coercive electric field limit, piezoelectric ceramics show hysteretic strain and electric flux responses, often referred to as minor loop hysteretic responses. The minor loop hysteretic response is frequency dependent and as a result piezoelectric composites also show frequency-dependent hysteretic behaviors. When cyclic electric fields with high amplitude, above the coercive electric field limit, are considered, piezoelectric ceramics experience polarization switching, which forms a major loop hysteresis. The polarization switching response is also time dependent. Furthermore, a polymeric matrix exhibits a viscoelastic response, which eventually affects the hysteretic response of piezoelectric composites. Only limited experimental tests are available on the piezoelectric composites with PZT fibers undergoing polarization switching.

In order to evaluate the overall properties and responses of piezoelectric composites under various loading histories, several micromechanics models have been formulated. Both linear and nonlinear responses have been considered, including limited studies on the hysteretic responses of piezocomposites. Examples of micromechanical models for piezoelectric composites are. It should be noted that the piezoelectric constitutive models employed in these studies are restricted to piezoelectric phenomena within a polarized state only, in absence of polarization switching. Only limited micromechanical models are available to describe the hysteretic responses of piezoelectric composites due to polarization switching. Their approaches merely considered rate-independent hysteretic constitutive relations for the ferroelectric constituents. Recently, Lin and Muliana extended their micromechanical models for particle and fiber reinforced piezoelectric composites to include the time-dependent polarization switching behaviors of the piezoelectric inclusions. In their models, the effect of viscoelastic polymeric matrix on the overall electromechanical response of composites is also studied. In the preceding micromechanics models, a simplified microstructural characteristic was considered in formulating the homogenized (effective) electromechanical response of the piezoelectric composites. For example, Dunn and considered a microstructural characteristic based on the Mori-Tanaka model, in which a single inclusion is embedded in an effective (homogenized) matrix medium, while considered a unit-cell model with several inclusion and matrix subcells. One of the main advantages of the models with a simplified microstructural characteristic is that it allows incorporating rigorous nonlinear constitutive models for the constituents, while being computationally efficient, with the scope of predicting the overall response of composites. However, this type of micromechanics model is limited in capturing the variations in the field variables (stress, strain, electric flux, electric field, displacement, electric potential), including the effect of localized and discontinuity of the field variables on the overall response of the composites. Another type of micromechanics model considers more detailed microstructural morphologies, such as distribution, size, and shape of the inclusions, possible existence of voids, placement of the electrodes, etc. These micromechanics models have an advantage in capturing the variations in the field variables including the localized and discontinuity in field variables.

Capacitor Dielectric and Piezoelectric Ceramics

Capacitor dielectric and piezoelectric ceramics are advanced industrial materials that, by virtue of their poor electrical conductivity, are useful in the production of electrical storage or generating devices.

Capacitors are devices that store electric energy in the form of an electric field generated in the space between two separated, oppositely charged electrodes. Their capacity to store energy makes them essential components in many electric circuits, and that capacity can be greatly increased by inserting a solid dielectric material into the space separating the electrodes. Dielectrics are materials that are poor conductors of electricity. The non-conducting properties of ceramics are well known, and some ceramics are made into extremely effective dielectrics. Indeed, more than 90 percent of all capacitors are produced with ceramic materials serving as the dielectric.

Piezoelectrics are materials that generate a voltage when they are subjected to mechanical pressure; conversely, when subjected to an electromagnetic field, they exhibit a change in dimension. Many piezoelectric devices are made of the same ceramic materials as capacitor dielectrics.

Ferroelectric Properties of Barium Titanate

Low electric conductivity is a factor of the chemical bonds that form a material. In dielectrics, unlike in conductive materials such as metals, the strong ionic and covalent bonds holding the atoms together do not leave electrons free to travel through the material under the influence of an electric field. Instead, the material becomes electrically polarized, its internal positive and negative charges separating somewhat and aligning parallel to the axis of the electric field. When employed in a capacitor, this polarization acts to reduce the strength of the electric field maintained between the electrodes, which in turn raises the amount of charge that can be stored.

Most ceramic capacitor dielectrics are made of barium titanate ($BaTiO_3$) and related perovskite compounds. Perovskite ceramics have a face-centred cubic (fcc) crystal structure. In the case of $BaTiO_3$, at high temperatures (above approximately 120 °C, or 250 °F) the crystal structure consists of a tetravalent titanium ion (Ti^{4+}) sitting at the centre of a cube with the oxygen ions (O^{2-}) on the faces and the divalent barium ions (Ba^{2+}) at the corners. Below 120 °C, however, a transition occurs. The Ba^{2+} and O^{2-} ions shift from their cubic positions, and the Ti^{4+} ion shifts away from the cube centre. A permanent dipole results, and the symmetry of the atomic structure is no longer cubic (all axes identical) but rather tetragonal (the vertical axis different from the two horizontal axes). There is a permanent concentration of positive and negative charges toward opposite poles of the vertical axis. This spontaneous polarization is known as ferroelectricity; the temperature below which the polarity is exhibited is called the Curie point. Ferroelectricity is the key to the utility of $BaTiO_3$ as a dielectric material.

Within local regions of a crystal or grain that is made up of these polarized structures, all the dipoles line up in what is referred to as a domain, but, with the crystalline material consisting of a multitude of randomly oriented domains, there is overall cancellation of the polarization. However, with the application of an electric field, as in a capacitor, the boundaries between adjacent domains can move, so that domains aligned with the field grow at the expense of out-of-alignment domains, thus producing large net polarizations. The susceptibility of these materials to electric polarization is directly related to their capacitance, or capacity to store electric charge. The capacitance of a specific dielectric material is given a measure known as the dielectric constant, which is essentially the ratio between the capacitance of that material and the capacitance of a vacuum. In the case of the perovskite ceramics, dielectric constants can be enormous—in the range of 1,000–5,000 for pure $BaTiO_3$ and up to 50,000 if the Ti^{4+} ion is replaced by zirconium (Zr^{4+}).

Chemical substitutions in the $BaTiO_3$ structure can alter a number of ferroelectric properties. For example, $BaTiO_3$ exhibits a large peak in dielectric constant near the Curie point—a property that is undesirable for stable capacitor applications. This problem may be addressed by the substitution of lead (Pb^{2+}) for Ba^{2+}, which increases the Curie point; by the substitution of strontium (Sr^{2+}), which lowers the Curie point; or by substituting Ba^{2+} with calcium (Ca^{2+}), which broadens the temperature range at which the peak occurs.

Disk, Multilayer and Tubular Capacitors

Barium titanate can be produced by mixing and firing barium carbonate and titanium dioxide, but liquid-mix techniques are increasingly used in order to achieve better mixing, precise control of the barium-titanium ratio, high purity, and submicrometre particle size. Processing of the resulting powder varies according to whether the capacitor is to be of the disk or multilayer type. Disks are dry-pressed or punched from tape and then fired at temperatures between 1,250° and 1,350 °C (2,280° and 2,460 °F). Silver-paste screen-printed electrodes are bonded to the surfaces at 750 °C (1,380 °F). Leads are soldered to the electrodes, and the disks are epoxy-coated or wax-impregnated for encapsulation.

The capacitance of ceramic disk capacitors can be increased by using thinner capacitors; unfortunately, fragility results. Multilayer capacitors (MLCs) overcome this problem by interleaving dielectric and electrode layers. The electrode layers are usually palladium or a palladium-silver alloy. These metals have a melting point that is higher than the sintering temperature of the ceramic, allowing the two materials to be cofired. By connecting alternate layers in parallel, large capacitances can be realized with the MLC. The dielectric layers are processed by tape casting or doctor blading and then drying. Layer thicknesses as small as 5 micrometres (0.00022 inch) have been achieved. Finished "builds" of dielectric and electrode layers are then diced into cubes and cofired. MLCs have the advantages of small size, low cost, and good performance at high frequencies, and they are suitable for surface mounting on circuit boards. They are increasingly used in place of disk capacitors in most electronic circuitry. Where monolithic units are still employed, tubular capacitors are often used in place of disks, because the axial wire lead configuration of tubular capacitors is preferred over the radial configuration of disk capacitors for automatic circuit-board insertion machines.

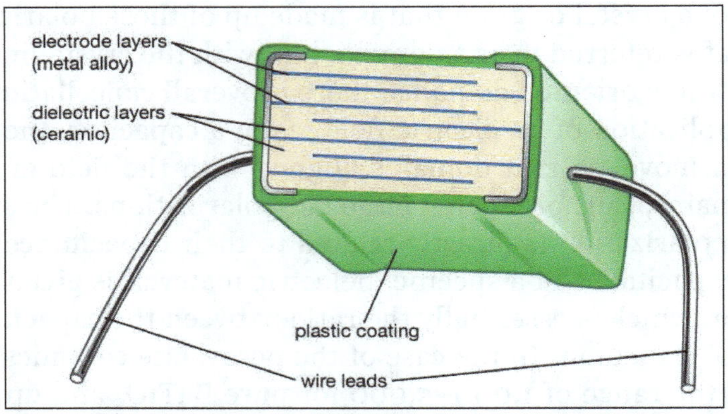

Schematic diagram of a multilayer capacitor, showing alternating layers of metal electrodes and ceramic dielectric.

Barium titanate-based MLCs usually require firing temperatures in excess of 1,250 °C. To facilitate cofiring with electrode alloys of lower melting temperatures, the sintering temperature of the ceramic can be reduced to the neighborhood of 1,100 °C (2,000 °F) by adding low-melting glasses or fluxing agents. In order to reduce the costs associated with precious-metal electrodes such as palladium and silver, ceramic compositions have been developed that can be cofired with less expensive nickel or copper at lower temperatures.

Barrier-layer Capacitors

Two other strategies to produce ceramic materials with high dielectric constants involve surface barrier layers or grain-boundary barrier layers; these are referred to as barrier-layer (BL) capacitors. In each case conductive films or grain cores are formed by donor doping or reduction firing of the ceramic. The surface or grain boundaries are then oxidized to produce thin resistive layers. In surface BL capacitors oxidation is accomplished by adding oxidizing agents such as manganese oxide or copper oxide to the silver electrode paste prior to firing. In grain-boundary BL capacitors slow cooling in air or oxygen allows oxygen to diffuse into the grain boundaries and reoxidize thin layers adjacent to the boundaries. Oxidizing agents such as bismuth and copper oxides also can be incorporated into the electrode paste to diffuse along grain boundaries during firing. In either case very high apparent dielectric constants, 50,000 to 100,000, can be obtained. Care must be taken in using BL capacitors, however, as they have very low dielectric breakdown strengths. Dielectric breakdown involves sudden failure of and catastrophic discharge through the dielectric material, with usually irreversible damage to the ceramic. In BL capacitors the barriers are so thin that local fields can be quite intense.

Random-access Memories

An extremely important application of thin-film ferroelectrics is in random-access memories (RAMs) for computers. Because of their larger dielectric constants, titanate-based ferroelectrics can achieve higher bit densities than silica-based semiconductors when used as thin-film capacitors in dynamic random-access memories (DRAMs). They also can be used as ferroelectric random-access memories (FERAMs), where the opposing directions of polarization can represent the two states of binary logic. Unlike conventional semiconductor RAM, the information stored in FERAMs is nonvolatile; *i.e.*, it is retained when the power is turned off.

Optical Ceramics

Optical ceramics are advanced industrial materials developed for use in optical applications. Optical materials derive their utility from their response to infrared, optical, and ultraviolet light. The most obvious optical materials are glasses, but ceramics also have been developed for a number of optical applications.

Passive Devices

Optical and Infrared Windows

In their pure state, most ceramics are wide-band-gap insulators. This means that there is a large gap of forbidden states between the energy of the highest filled electron levels and the energy of the next highest unoccupied level. If this band gap is larger than optical light energies, these ceramics will be optically transparent (although powders and porous compacts of such ceramics will be white and opaque due to light scattering). Two applications of optically transparent ceramics are windows for bar-code readers at supermarkets and infrared radome and laser windows.

Sapphire (a single-crystal form of aluminum oxide, Al_2O_3) has been used for supermarket checkout windows. It combines optical transparency with high scratch resistance. Similarly, single-crystal or infrared-transparent polycrystalline ceramics such as sodium chloride (NaCl), rubidium-doped potassium chloride (KCl), calcium fluoride (CaF), and strontium fluoride (SrF_2) have been used for erosion-resistant infrared radomes, windows for infrared detectors, and infrared laser windows. These polycrystalline halide materials tend to transmit lower wavelengths than oxides, extending down to the infrared region; however, their grain boundaries and porosity scatter radiation. Therefore, they are best used as single crystals. As such, however, halides are insufficiently strong for large windows: they can plastically deform under their own weight. In order to strengthen them, single crystals are typically hot-forged to induce clean grain boundaries and large grain sizes, which do not decrease infrared transmission significantly but allow the body to resist deformation. Alternatively, large-grained material can be fusion-cast.

Lamp Envelopes

Electric discharge lamps, in which enclosed gases are energized by an applied voltage and thereby made to glow, are extremely efficient light sources, but the heat and corrosion involved in their operation push optical ceramics to their thermochemical limits. A major breakthrough occurred in 1961, when Robert Coble of the General Electric Company in the United States demonstrated that alumina (a synthetic polycrystalline, Al_2O_3) could be sintered to optical density and translucency using magnesia (magnesium oxide, MgO) as a sintering aid. This technology permitted the extremely hot sodium discharge in the high-pressure sodium-vapour lamp to be contained in a refractory material that also transmitted its light. The plasma within the inner alumina lamp envelope reaches temperatures of 1,200 °C (2,200 °F). Energy emission covers almost the entire visible spectrum, creating a bright white light that reflects all colors—unlike that of the low-pressure sodium-vapour lamp, whose amber glow is common in the skylines of major cities.

Pigments

The ceramic color or pigment industry is a long-standing, traditional industry. Ceramic pigments or stains are made of oxide or selenide compounds in combination with specific transition-metal or rare-earth elements. Absorption of certain wavelengths of light by these species imparts specific colors to the compound. For example, cobalt aluminate ($CoAl_2O_4$) and cobalt silicate (Co_2SiO_4) are blue; tin-vanadium oxide (known as V-doped SnO_2) and zirconium-vanadium oxide (V-doped ZrO_2) are yellow; cobalt chromite ($CoCr_2O_3$) and chromium garnet ($2CaO \cdot Cr_2O_3 \cdot 3SiO_2$) are green; and chromium hematite ($CrFe_2O_3$) is black. A true red color, unavailable in naturally occurring silicate materials, is found in solid solutions of cadmium sulfide and cadmium selenide (CdS-CdSe).

Powdered pigments are incorporated into ceramic bodies or glazes in order to impart color to the fired ware. Thermal stability and chemical inertness during firing are important considerations.

Active Devices

Phosphors

Ceramic phosphors are employed for both general lighting (as in fluorescent lights) and for electronic imaging (as in cathode-ray tubes). Phosphors function when electrons within them are stimulated from stable, low-energy positions to higher levels by an appropriate means—e.g., thermal, optical, X-ray, or electron excitation. When the energized electrons drop back to lower energy levels, light can be emitted at one or more characteristic wavelengths. These wavelengths are determined by controlled dopants, referred to as activators. Examples of activated phosphors (and their resulting color emissions) are lead-activated calcium tungstate (blue), manganese-activated zircon (green), lead- or manganese-activated calcium silicate (yellow to orange), and europium-activated yttrium vanadate (red). There are countless other examples.

Two major applications of phosphor ceramics are in cathode-ray tubes (CRTs) for television sets and computer monitors. Thin layers of phosphor powders are applied to the inside of the display screen of the CRT. Electrons are accelerated from the cathode toward the screen, directed by magnetic coils. Light emission (phosphorescence) occurs wherever the electron beam strikes the phosphor layer, and images are formed by high-speed scanning of the electron beam over the surface of the screen. Color screens employ interspersed small dots of phosphors of each of the three primary colors (red, yellow, and blue), with separate electron beams to address each color.

Efficient indoor lighting is usually accomplished by fluorescent lamps. Phosphors of a suitably doped calcium halophosphate are deposited as thin powder layers on the inner surfaces of thin-walled glass tubes. The tubes are evacuated and backfilled with a mixture of mercury vapour and an inert gas. An electric discharge through the gas causes the mercury vapour to emit energy in the ultraviolet range, which strikes the phosphor layer and stimulates visible light emission. The resulting combination of blue and orange emission is comparable to that of incandescent lamps. Phosphors must be manufactured by clean-room methods in order to eliminate unwanted impurities that can "kill" phosphorescence.

Lasers

Lasing, or "light amplification by stimulated emission of radiation," takes place in various media, including glasses and single-crystal ceramics. The first laser, operated by Theodore H. Maiman in 1960, consisted of a rod of synthetic ruby (single-crystal Al_2O_3 doped with chromium) that was excited by a flash lamp. Excitation, or pumping, involves promoting electrons within the dopant centres to higher energy levels by optical or electronic means. The decay of the stimulated electron to a lower energy state yields emission of light, which is contained within the lasing solid between two mirrors (one completely silvered and one partially silvered). As the emitted light reflects back and forth, it stimulates other centres until an intense, coherent, narrow beam of monochromatic light is released. Two well-known ceramic lasing materials are the chromium-doped Al_2O_3 known as ruby and a neodymium-doped yttrium aluminum garnet known as Nd-YAG.

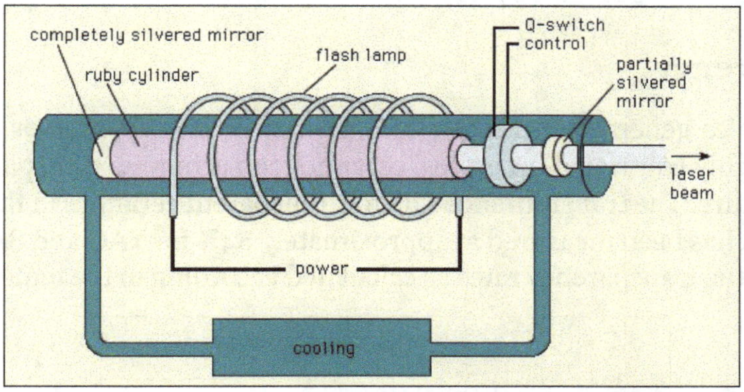

Ruby laser being used in a Q-switch, a special switching device that produces giant output pulse.

Electro-optical Components

Electro-optical ceramics are materials that combine optical transparency with voltage-variable optical, or electro-optical (EO), behavior. Single-crystal EO materials include lithium niobate ($LiNbO_3$) and lithium tantalate ($LiTaO_3$); polycrystalline EO materials include a lanthanum-modified lead zirconate tantalate known as PLZT. Among other EO properties, these materials exhibit voltage-dependent birefringence. Birefringence is the difference between the refractive index parallel to the optical axis of the crystal and the refractive index perpendicular to the optical axis. Because the propagation velocity is different in the two directions, a phase shift occurs, and this phase shift can be varied by an applied voltage. Such EO behavior is the basis of a number of optical devices, including switches, modulators, and demodulators for high-speed optical communications. EO ceramic thin films also can be integrated with silicon semiconductors in so-called optoelectronic integrated circuits (OEICs).

Transparent Ceramics

Transparent ceramics are an optical material that combines the high performance of single crystal with the productivity of glass and resin. Transparent ceramics have a higher thermal resistance than glass or resin and they are stronger and harder. They can be shaped more freely and their productivity is more efficient than single crystal. At present, we manufacture the transparent ceramics YAG ($Y_3Al_5O_{12}$/yttrium aluminum garnet) and Y_2O_3 (yttrium oxide). Furthermore, through the use of doping, we can add optical functions to transparent ceramics such as wavelength conversion.

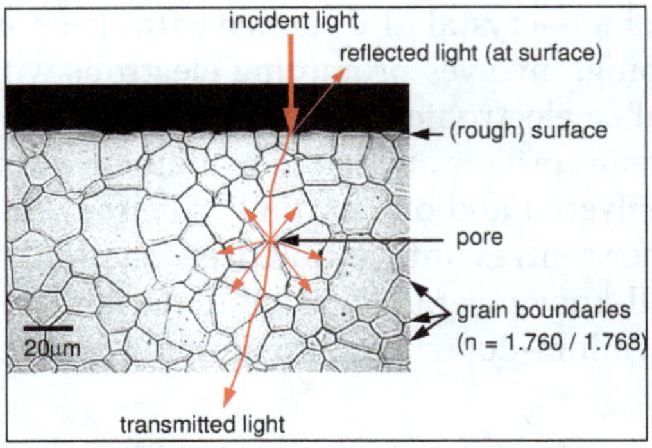

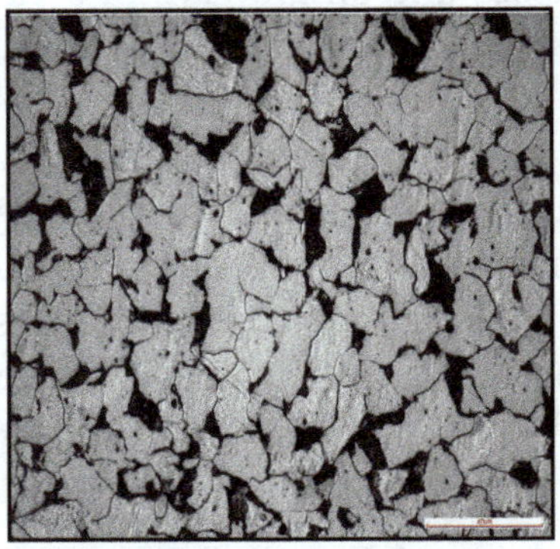

Optical Transparency

Polycrystalline ceramics generally consist of opaque matter that absorbs or scatters visible light. However, through highly controlled processes, ceramics can achieve a transparency similar to glass or single crystal sapphire. The transmittance rate for transparent ceramics in the visible light to middle infrared spectrum has been measured at approximately 84% for YAG and 81% for Y_2O_3. These ratios are equivalent to the transparency rates as calculated according to the materials refractive index.

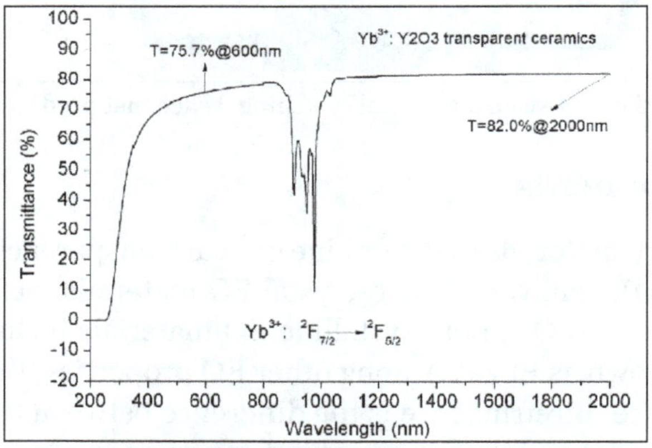

Transmittance of Transparent Ceramics.

Thermal Resistance, Strength and Hardness

High temperature sintered transparent ceramics have a higher resistance and are stronger and harder than optical glass or resin. They maintain their transparency at temperatures of up to 1200 °C, and they can be used stably in oxidized environments.

Resistance to Chemicals and Corrosion

Highly pure, highly dense transparent ceramics hold up extremely well against oxygen and alkalis, and they resist highly corrosive halogen plasma. They can be used in harsh environments that glass and resins cannot tolerate.

Optical Functions

Through doping, it is possible to add optical functions such as wavelength conversion to transparent ceramics. For example, laser oscillation has been successfully achieved by doping YAG and Y_2O_3 transparent ceramics with neodymium (Nd) or ytterbium (Yb), the results of which reveal the same characteristics as single crystal.

Mass Productivity

Producing a single crystal optical part generally requires a considerable amount of time and effort since it must be cut from an ingot. This involves a fair amount of material loss during the machining process. On the other hand, with ceramics, it is possible to create and sinter a part from material that has been manufactured close to the shape of the final product, making this method suitable for low volume sample manufacturing and multi-variety manufacturing as well as high volume mass production.

Table: Material Properties.

Material	Transparent ceramics		Optical glass	Single crystal	Optical resin
Property	YAG	Y_2O_3	BK-7	Sapphire	PMMA
Density (g/cm^3)	4.54	5.03	2.52	3.98	1.19
Flexural strength (MPa)	247	219	70	690	100
Young's modulus (GPa)	283	175	78.6	470	-
Poisson ratio	0.25	0.30	0.22	0.23	-
Vickers hardness (GPa)	11.3	7.0	5.8	22.5	0.16
Thermal expansion ($x\ 10^{-6}/K$)	9.2	8.7	7.0	7.0	70
Thermal conductivity (Wink)	10	11	1.2	42	0.2

Table: Refractive Index and Abbe Number.

Wavelength		Transparent ceramics	
		YAG	$Y^{2}0^{3}$
C Line	656.3 nm	1.827	1.909
D Line	587.6 nm	1.832	1.916
E Line	546.1 nm	1.836	1.922
F Line	486.1 nm	1.843	1.943
G Line	435.8 nm	1.852	1.949
Abbe number		52.5	36.2

Technical Description

Ceramics are produced by sintering powdered materials that generally have light scattering characteristics. This is attributable to the ceramic powder having spaces between each grain of powder, the inconsistencies among the grains and the impurities that are present at the boundaries of the grains. Using a highly controlled process, we have been able to remove the light scattering attributes to develop a transparent ceramic, YAG ($Y_3A_{15}O_{12}$; yttrium aluminum garnet) and Y_2O_3 (yttrium oxide). These transparent ceramics have a polycrystalline structure with a transmittance rate that has been measured in the visible light to middle infrared spectrum at approximately 84% for YAG and 81% for Y_2O_3, which is equivalent to the transparency ratios calculated according to the materials refractive index.

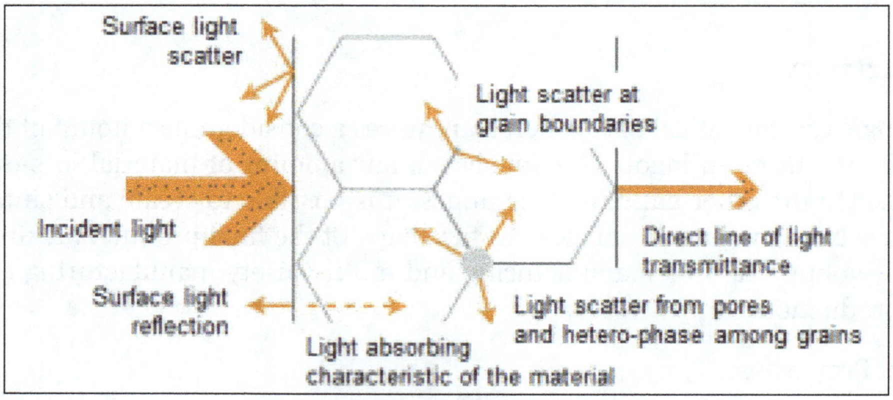

Causes of light scatter in ceramics.

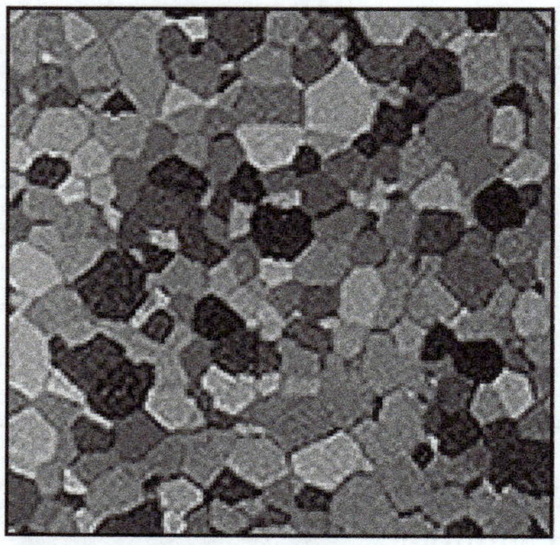

Microstructure of Transparent ceramics.

Application Examples: Lenses, windows, laser hosts, scintillators, etc.

Conductive Ceramics

Conductive ceramics, advanced industrial materials that, owing to modifications in their structure, serve as electrical conductors.

In addition to the well-known physical properties of ceramic materials—hardness, compressive strength, brittleness—there is the property of electric resistivity. Most ceramics resist the flow of electric current, and for this reason ceramic materials such as porcelain have traditionally been made into electric insulators. Some ceramics, however, are excellent conductors of electricity. Most of these conductors are advanced ceramics, modern materials whose properties are modified through precise control over their fabrication from powders into products.

Electric conductivity in ceramics, as in most materials, is of two types: electronic and ionic. Electronic conduction is the passage of free electrons through a material. In ceramics the ionic bonds holding the atoms together do not allow for free electrons. However, in some cases impurities of differing valence (that is, possessing different numbers of bonding electrons) may be included in the material, and these impurities may act as donors or acceptors of electrons. In other cases transition metals or rare-earth elements of varying valency may be included; these impurities may act as centres for polarons—species of electrons that create small regions of local polarization as they move from atom to atom. Electronically conductive ceramics are used as resistors, electrodes, and heating elements.

Ionic conduction consists of the transit of ions (atoms of positive or negative charge) from one site to another via point defects called vacancies in the crystal lattice. At normal ambient temperatures very little ion hopping takes place, since the atoms are at relatively low energy states. At high temperatures, however, vacancies become mobile, and certain ceramics exhibit what is known as fast ionic conduction. These ceramics are especially useful in gas sensors, fuel cells, and batteries.

Thick-film, Thin-film Resistors and Electrodes

Semimetallic ceramic conductors have the highest conductivities of all but superconducting ceramics. Examples of semimetallic ceramics are lead oxide (PbO), ruthenium dioxide (RuO_2), bismuth ruthenate ($Bi_2Ru_2O_7$), and bismuth iridate ($Bi_2Ir_2O_7$). Like metals, these materials have overlapping electron energy bands and are therefore excellent electronic conductors. They are used as "inks" for screen printing resistors into thick-film microcircuits. Inks are pulverized conductor and glaze particles dispersed in suitable organics, which impart the flow properties necessary for screen printing. On firing, the organics burn out as the glazes fuse. By varying the amount of conductor particles, it is possible to produce wide variations in the resistance of thick films.

Ceramics based upon mixtures of indium oxide (In_2O_3) and tin oxide (SnO_2)—referred to in the electronics industry as indium tin oxide (ITO)—are outstanding electronic conductors, and they have the added virtue of being optically transparent. Conductivity and transparency arise from the combination of a large band gap and the incorporation of sufficient electron donors. There is thus an optimal electron concentration to maximize both electronic conductivity and optical transmission. ITO sees extensive application as thin transparent electrodes for solar cells and for liquid-crystal displays such as those employed in laptop computer screens. ITO also is employed as a thin-film resistor in integrated circuits. For these applications it is applied by standard thin-film deposition and photolithographic techniques.

Heating Elements

A longstanding use of conductive ceramics is as heating elements for electric heaters and electrically heated furnaces. Conductive ceramics are especially effective at elevated temperatures and in oxidizing environments where oxidation-resistant metal alloys fail. Examples of electrode ceramics and their temperatures of maximum use in air are shown in table. Each material has a unique conduction mechanism. Silicon carbide (SiC) normally is a semiconductor; suitably doped, however, it is a good conductor. Both SiC and molybdenum disilicide ($MoSi_2$) form protective silica-glass surface layers, which protect them from oxidation in oxidizing atmospheres. $MoSi_2$ is a semimetal with a high conductivity. Lanthanum chromite ($LaCr_2O_4$) is a small polaron conductor;

substituting alkaline-earth ions (*e.g.*, calcium, or Ca^{2+}) for La^{3+} results in an equal proportion of Cr^{3+} being converted to Cr^{4+}. Hopping of electrons between the two states of Cr ions yields high conductivity, especially at elevated temperatures.

Heating element ceramics		
Ceramic material		Temperature of maximum use in air
Common name	Chemical formula	(°C/°F)
Silicon carbide	Sic	1,500/2,730
Molybdenum disilicide	$Mosi_2$	1,800/3,270
Lanthanum chromite	$Lacr_2o_4$	1,800/3,270
Zirconia	Zro_2	2,200/3,630

Conduction in zirconia (ZrO_2) is ionic, as opposed to the electronic conduction mechanisms described above. When zirconia is doped with Ca^{2+} or yttrium ions (Y^{3+}), oxygen vacancies are produced. Above 600 °C (1,100 °F), oxygen ions (O^{2-}) become mobile and fill these vacancies, and they are highly mobile at higher temperatures. Zirconia heating elements require a preheater to reach the 600 °C threshold, but they can be used to achieve temperatures up to 2,000° C (3,600 °F).

Tin oxide (SnO_2) has a very specific application as the preferred electrode for specialty glass-melting furnaces (as for optical glass). This application requires high conductivity and resistance to the corrosive elements in glass melts; in addition, corroded electrode material must not discolor the glass. Tin oxide is the only material that satisfies these criteria. Pure tin oxide is a wide band-gap semiconductor, but inherent oxygen deficiency plus the substitution of antimony ions for tin result in high conductivity.

Thermistors

Thermistors, or thermally sensitive resistors, are electric resistors whose resistive properties vary with temperature. They are made of materials that have high temperature coefficients of resistance (TCR), the value that describes resistance change with temperature. Negative TCR, or NTCR, ceramics are materials whose electric resistance decreases as temperatures rise. These ceramics are usually spinels based on oxides of iron, cobalt, and manganese that exhibit small polaron conduction. Under normal temperatures there is an energy barrier to moving electrons from site to site. As thermal energy rises with temperature, however, the ability of electrons to surmount this barrier increases, so that resistivity goes down—hence the NTCR behavior. Extensive solid solutions are possible in these materials (that is, a large number of foreign ions can substitute for the host ions in the crystal structure), so that the resistances and temperature coefficients can be tailored over wide ranges.

NTCR thermistors are used as temperature sensors and as temperature compensation resistors. The beam focus coil in cathode-ray tubes for televisions and computers relies on NTCR thermistors to compensate for the resistivity of the coil material. Thermistors also are used as fuel-level sensors in gas tanks. When a thermistor under constant voltage is immersed in fuel, it loses more heat than when it is surrounded only by vapour. The difference in heat loss results in a change in resistance, which in turn changes the flow of current through the fuel sensor.

Gas Sensors

Carbon Monoxide Sensors

In addition to the heating electrode applications noted above, tin oxide also is used in carbon monoxide gas sensors for home and industry. Adsorption of carbon monoxide at contacts between particles of SnO_2 produces local charge states that alter the electric properties (*e.g.*, resistance, capacitance) of the porous, polycrystalline material. When life-threatening concentrations of carbon monoxide are detected, an alarm is triggered. By changing the temperature of operation, the sensor can be made selective for a variety of reducing gas species (such as hydrogen, carbon monoxide, and hydrocarbons).

Oxygen Sensors

Oxygen sensors are employed in industry to monitor and control processing atmospheres and also in automobiles to monitor and control the air-to-fuel (A/F) ratio in the internal combustion engine. A prominent sensor material is zirconia, which, as noted above, can be an excellent high-temperature oxygen conductor if suitably doped with Ca^{2+} or Y^{3+}. A tube or thimble made of zirconia can be exposed on its exterior to the hot atmosphere to be monitored and on its interior to air, with high-temperature seals preventing leakage between the two environments. Porous platinum electrodes on the two surfaces can be used to register a galvanic cell voltage across the solid zirconia electrolyte that is proportional to the difference in oxygen content between the exterior atmosphere and the interior air.

Each automobile has a zirconia oxygen sensor such as that illustrated in figure inserted into its hot exhaust manifold. The primary function of the oxygen sensor there is to control the A/F ratio through appropriate feedback circuitry to the fuel injection system. Control is necessary to protect the catalytic converter elements from being poisoned at A/F ratios that are too high or too low.

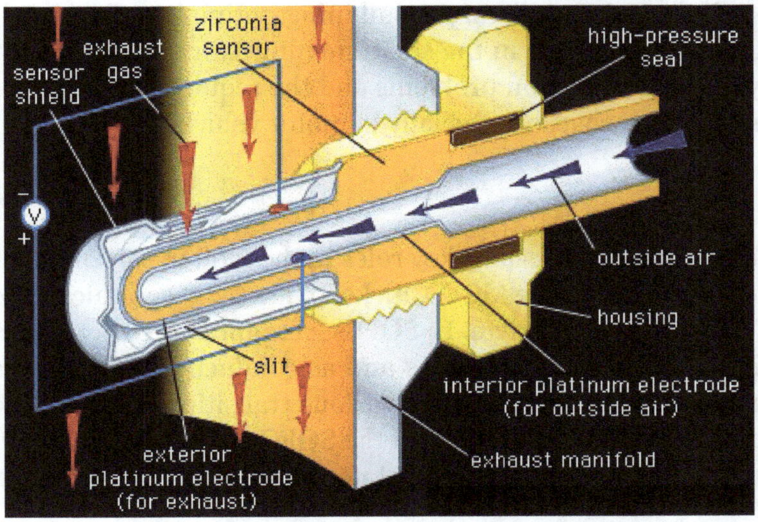

Figure shows schematic diagram of a zirconia oxygen sensor used to monitor automobile exhaust gases. The sensor, approximately the size of a spark plug, is fitted into the exhaust manifold of an automobile engine. The thimble-shaped zirconia sensor, sandwiched between thin layers of porous platinum, is exposed on its interior to outside air and on its exterior to exhaust gas passing through slits in the sensor shield. The two platinum surfaces serve as electrodes, conducting a voltage across the zirconia that varies according to the difference in oxygen content between the exhaust gas and the outside air.

Batteries and Fuel Cells

Two other galvanic applications of conductive ceramics are in batteries and fuel cells. A battery is a device that converts chemical energy into electricity. In its simplest form it consists of two metal or metal oxide elements, called the anode and the cathode, immersed in a liquid or solid chemical compound called the electrolyte. Ion flow in the electrolyte is accompanied by a compensating movement of electrons from the anode; the electrons flow through an appropriate conductor to the cathode, and the electric circuit is complete. Batteries are ubiquitous in modern life, finding use in toys, portable appliances, and motor vehicles.

Fuel cells produce electric power like a battery, except that power production is prolonged by supplying a gaseous or liquid fuel to the anode and air or oxygen to the cathode. Fuel cells have been developed for load-leveling in electric power plants, but they also may be employed in motor vehicles.

Batteries

High-energy-density batteries based on sodium beta-alumina have been developed for vehicular applications. Beta-alumina has the ideal formula $Na_2O{\cdot}11Al_2O_3$. It has a complicated structure consisting of spinel blocks sandwiching conduction planes in which sodium cations (Na^+) can rapidly migrate. It is therefore known as a fast sodium ion conductor. A related structure is beta″-alumina, $Na_2MgAl_{10}O_{17}$, where magnesium cations (Mg^{2+}) stabilize the structure and require additional Na^+ in the conduction plane for charge compensation. These materials must be carefully processed in order to achieve uniform microstructures combining optimal strength and ionic conductivity. They are used as the solid electrolyte in the sodium-sulfur storage battery. Although this battery exhibits high energy density, corrosion problems and the requirement that the battery operate at elevated temperatures are drawbacks, especially in motor vehicles.

Fuel Cells

Of the several fuel cell types, ceramics play key roles in the molten carbonate fuel cell (MCFC) and the solid oxide fuel cell (SOFC). In the MCFC, nickel oxide (NO) ceramics serve as porous anodes for the molten salt (carbonate) electrolyte. In SOFCs, ceramics serve not only as the solid electrolyte (in this case, zirconia) but also as anodes and as conductive connections between adjacent cells (cobaltites, manganites, and chromites of various transition metals). Anode materials must be excellent electronic conductors. In the case of the SOFC anode, conductivity is accomplished by small polaron conduction between two valence states of the transition metal constituent.

The processing of SOFCs is an extremely difficult proposition. Anode, electrolyte, cathode, and interconnecting layers must be suitable for firing together. Thermal expansion mismatch therefore must be minimized. Parts of the structure must be dense and gas-impervious (the electrolyte), whereas others are made intentionally porous (the electrodes).

Superconductors

Superconductivity is the complete disappearance of electric resistance in materials that are cooled to extremely low temperatures. The temperature at which resistance ceases is referred to as the transition temperature, or critical temperature (T_c). T_c is usually measured in degrees kelvin (K)—0 K being absolute zero, the temperature at which all atomic motion ceases. The best ceramic conductors are the so-called high T_c superconductors, materials that lose their resistance at much higher critical temperatures than their metal alloy counterparts. Most high T_c ceramics are layered structures, with two-dimensional copper-oxygen sheets along which superconduction takes place. The first of these was discovered in 1986 by the Swiss researchers J. Georg Bednorz and Karl Alex Müller. Within a year an yttrium barium copper oxide ceramic, $YBa_2Cu_3O_7$, had been discovered to have a T_c higher than 77 K, the boiling point of nitrogen (−195.8 °C, or −320.4 °F). This finding raised the possibility of practical superconductors being cooled by liquid nitrogen—as opposed to conventional superconducting materials, which have to be cooled by more expensive liquid helium.

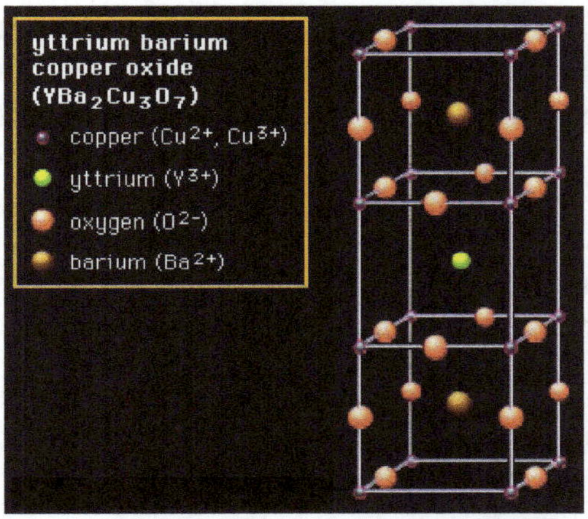

The arrangement of copper, yttrium, oxygen, and barium ions in yttrium barium copper oxide ($YBa_2Cu_3O_7$); an example of a superconducting ceramic crystal structure.

Although still higher transition temperatures have since been achieved, ceramic superconductors are difficult to process (in contrast to metal alloy superconductors), and they are notoriously brittle—properties that have limited their application. In hospitals and clinics small superconducting magnets are used in magnetic resonance imaging (MRI) apparatuses, where they generate the large magnetic fields necessary to excite and then image atomic nuclei in body tissues. Potential applications include wires for highly efficient superconducting magnets and low-loss electric power transmission lines, as well as advanced devices such as Josephson junctions and so-called SQUIDs (superconducting quantum interference devices). Josephson junctions, formed at contacts between two superconductors, can convert a direct voltage into an alternating current whose frequency rises with applied voltage. Frequencies in the superhigh frequency (SHF) range can be achieved. SQUIDs are highly sensitive magnetic-field sensors based on a superconducting ring with a weak link, a point where the material reverts to its normal, non-superconducting state at a small current relative to the rest of the ring. SQUIDs are widely used in geophysics for measuring magnetic field oscillations of the Earth. They also are used to record magnetograms of organs in the human body. Conductive ceramics are only one of several types of electroceramics.

Magnetic Ceramics

Magnetic ceramics are oxide materials that exhibit a certain type of permanent magnetization called ferrimagnetism. Commercially prepared magnetic ceramics are used in a variety of permanent magnet, transformer, telecommunications, and information recording applications.

Ferrites: Composition, Structure and Properties

Magnetic ceramics are made of ferrites, which are crystalline minerals composed of iron oxide in combination with some other metal. They are given the general chemical formula $M(Fe_xO_y)$, M representing other metallic elements than iron. The most familiar ferrite is magnetite, a naturally occurring ferrous ferrite ($Fe[Fe_2O_4]$ or Fe_3O_4) commonly known as lodestone. The magnetic properties of magnetite have been exploited in compasses since ancient times.

The magnetic behavior exhibited by the ferrites is called ferrimagnetism; it is quite different from the magnetization (called ferromagnetism) that is exhibited by metallic materials such as iron. In ferromagnetism there is only one kind of lattice site, and unpaired electron "spins" (the motions of electrons that cause a magnetic field) line up in one direction within a given domain. In ferrimagnetism, on the other hand, there is more than one kind of lattice site, and electron spins align so as to oppose one another—some being "spin-up" and some being "spin-down"—within a given domain. Incomplete cancellation of opposing spins leads to a net polarization, which, though somewhat weaker than for ferromagnetic materials, can be quite strong.

Three basic classes of ferrites are made into magnetic ceramic products. Based upon their crystal structure, they are the spinels, the hexagonal ferrites, and the garnets.

Spinels

Spinels have the formula $M(Fe_2O_4)$, where M is usually a divalent cation such as manganese (Mn^{2+}), nickel (Ni^{2+}), cobalt (Co^{2+}), zinc (Zn^{2+}), copper (Cu^{2+}), or magnesium (Mg^{2+}). M can also represent the monovalent lithium cation (Li^+) or even vacancies, as long as these absences of positive charge are compensated for by additional trivalent iron cations (Fe^{3+}). The oxygen anions (O^{2-}) adopt a close-packed cubic crystal structure, and the metal cations occupy the interstices in an unusual two-lattice arrangement. In each unit cell, containing 32 oxygen anions, 8 cations are coordinated by 4 oxygens (tetrahedral sites), and 16 cations are coordinated by 6 oxygens (octahedral sites). The antiparallel alignment and incomplete cancellation of magnetic spins between the two sublattices leads to a permanent magnetic moment. Because spinels are cubic in structure, with no preferred direction of magnetization, they are "soft" magnetically; *i.e.*, it is relatively easy to change the direction of magnetization through the application of an external magnetic field.

Hexagonal Ferrites

The so-called hexagonal ferrites have the formula $M(Fe_{12}O_{19})$, where M is usually barium (Ba), strontium (Sr), or lead (Pb). The crystal structure is complex, but it can be described as hexagonal with a unique *c* axis, or vertical axis. This is the easy axis of magnetization in the basic structure.

Because the direction of magnetization cannot be changed easily to another axis, hexagonal ferrites are referred to as "hard."

Garnet Ferrites

Garnet ferrites have the structure of the silicate mineral garnet and the chemical formula $M_3(Fe_5O_{12})$, where M is yttrium or a rare-earth ion. In addition to tetrahedral and octahedral sites, such as those seen in spinels, garnets have dodecahedral (12-coordinated) sites. The net ferrimagnetism is thus a complex result of antiparallel spin alignment among the three types of sites. Garnets are also magnetically hard.

Processing of Ceramic Ferrites

Ceramic ferrites are made by traditional mixing, calcining, pressing, firing, and finishing steps. Control of cation composition and gas atmosphere is essential. For example, the saturation magnetization of spinel ferrites can be greatly enhanced by partial substitution of $Zn(Fe_2O_4)$ for $Ni(Fe_2O_4)$ or $Mn(Fe_2O_4)$. The zinc cations prefer tetrahedral coordination and force additional Fe^{3+} onto the octahedral sites. This results in less cancellation of spins and greater saturation magnetization.

Advanced processing also is used for ferrite manufacture, including coprecipitation, freeze-drying, spray roasting, and sol-gel processing. In addition, single crystals are grown by pulling from fluxed melts (the Czochralski method) or by gradient cooling of melts (the Bridgman method). Ferrites also can be deposited as thin films on suitable substrates by chemical vapour deposition (CVD), liquid-phase epitaxy (LPE), and sputtering.

Applications

Permanent Magnets

Hard magnetic ferrites are used as permanent magnets and in refrigerator seal gaskets. They also are used in microphones and speaker gaskets. The largest market for permanent magnets is in small motors for cordless appliances and in automobile applications.

Electric Circuits

Although ceramic ferrites have smaller saturation magnetizations than magnetic metals, they can be made much more resistive to electric currents. In soft ferrites this is accomplished by oxidizing the grain boundaries during processing to yield resistive internal layers. Greater resistivity reduces the occurrence of eddy currents, which are caused by rapidly fluctuating magnetic fields and result in the loss of signal energy to heat. As a consequence of their reduction of eddy currents, soft ceramic ferrites are employed in telecommunications and transformers, especially at higher frequencies.

One of the earliest uses of ceramic ferrites was as a channel filter in telephony, although solid-state filters and fibre optics are making this application obsolete. Ferrites are used in the tone-generating circuitry of push-button telephones and as load coils in transmission lines to reduce signal loss over long distances. A major application is as deflection yokes and fly-back transformers for control of electron-beam rastering in television sets. Ferrites also are used extensively as input and output filters, switching regulators, and high-frequency power transformers in electric power systems.

Magnetic Storage

The earliest nonvolatile memory for computers consisted of tiny wire-threaded ferrite torroids; their two stable states of remanence were able to represent the "1" and "0" of binary logic. This application has long since been taken over by semiconductor memory devices, but ferrites still dominate external memory storage on devices such as tape, floppy disks, and hard disks. In place of torroids, very small regions of acicular (needle-like) particles adhering to the tape or disk can be magnetized by a field in the gap of a recording head. The head is also often made from ferrite material. Ferrite tape media, typically gamma-ferrite (γ-Fe_2O_3), is used for analog audio and video recording.

Computer bubble memory has been developed based on magnetic garnet ferrites. In bubble memory small, cylindrically reversed magnetic domains can be generated, moved, and stored in specified locations to be read at a later time. The presence or absence of a bubble corresponds to the two binary logic states.

Microwave and Radar

At microwave frequencies (from 1 to 300 gigahertz) ferrites offer gyromagnetic properties; *i.e.,* they can rotate and direct microwave energy in devices called circulators. Radar-absorbing paint made from ferrites can be used to coat military aircraft for stealth operations. Magnetic ceramics are only one of several types of electroceramics.

Bioceramics

Bioceramics are ceramic materials specially developed for use as medical and dental implants. They are usually used to replace hard tissue in the body like bone and teeth. Common bioceramics are alumina, zirconia and a form of calcium phosphate known as hydroxyapatite.

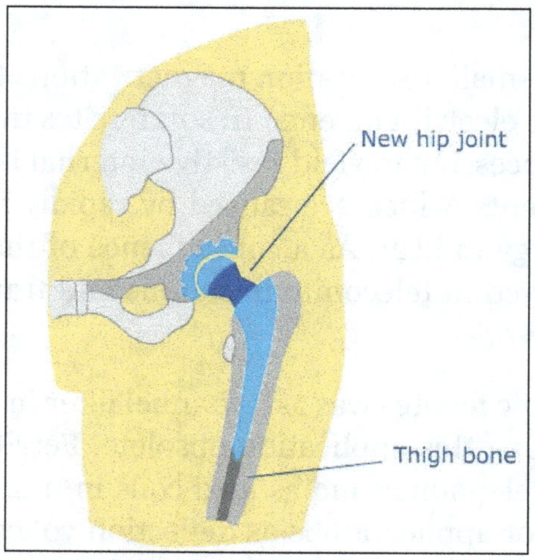

Hip Joint

Bioactive and Bioinert

Bioactive ceramics interact with the body so that tissue bonding and eventual incorporation into the body occurs after a time. Calcium phosphate-based bioceramics are bioactive. Bioinert ceramics do not interact with the body's environment apart from causing an initial 'fibrous tissue' reaction, which coats the ceramic. Alumina and zirconia-based ceramics are classed as bioinert.

Raw Materials

Alumina (Al_2O_3) is a white powder. When shaped, compressed and heated to a high temperature (sintering), the ceramic that results has high density, high strength, excellent corrosion resistance, good biocompatibility and high wear resistance. In addition, it can be machined, ground and polished to a high-quality product.

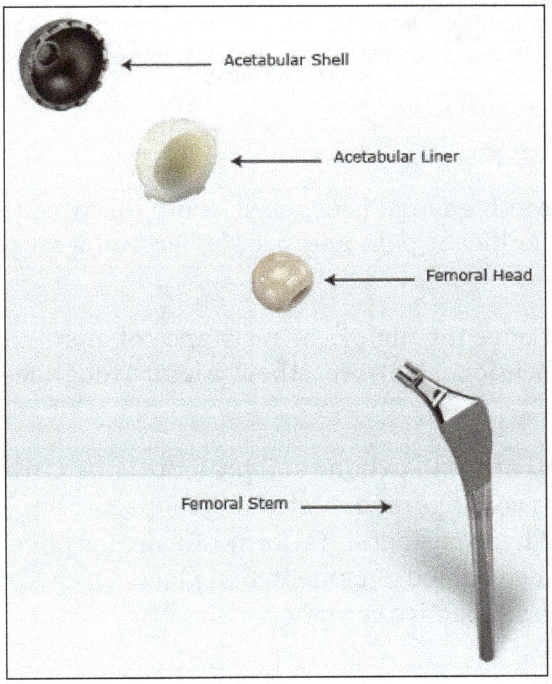

Hip replacement components.

Zirconia (ZrO_2) is also a white powder. Like alumina, it can be compressed and sintered into a very strong ceramic. Unlike alumina, its wear-resistance properties are not as good. By adding yttrium oxide and small amounts of magnesium oxide, a better-wearing bioceramic called Y-TZP can be made.

Calcium phosphate or hydroxyapatite – $Ca_{10}(PO_{46}(OH)_2$ – is the principal component of natural bone in the body. Ceramics made from synthetic calcium phosphate can also be used in medical applications. The problem is that these ceramics are not as strong as alumina or zirconia ceramics.

Bioceramic Bones and Teeth

Alumina bioceramics are used as replacement parts in hip and knee operations. The inertness of the ceramic, its high wear resistance and its excellent biocompatibility make it the ceramic of choice. The high load-bearing properties of alumina also makes it an ideal ceramic for dental implants.

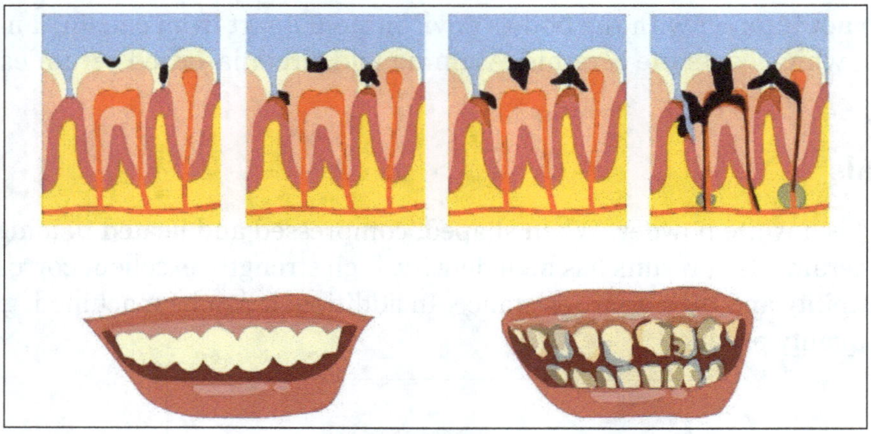

Tooth decay.

Tooth Decay

Once the dentin under the tooth enamel is exposed, tooth decay can set in. Over time, the decay can progress, resulting in significant pain and people becoming more self-conscious about their smiles.

Zirconia bioceramics do not have the high wear resistance of alumina bioceramics and are not as widely used in hip joint applications. However, their fracture toughness and bending strength give them additional qualities.

Calcium phosphate ceramics can bond to bone and promote bone growth at their surfaces. A popular use of these ceramics is as coatings on dental and orthopaedic implants. For example, titanium tooth root pegs coated with hydroxyapatite (a form of calcium phosphate) give a longer lasting implant than pegs that have been glued or cemented in place. The hydroxyapatite binds chemically with living bone because it is a bioactive ceramic.

Bovine Hydroxyapatite

Often as a result of injuries sustained in road crash accidents, bone reconstruction operations are needed. A novel method involving the use of bovine hydroxyapatite has been recently successfully tested in sheep and dogs.

Spongy bone material from cow bones is cut out and subjected to chemical and physical processes that remove all of the fat and protein. The remaining material is then heated to 1,000 °C. After this heating process, what remains is an open 3D mineral shell of hydroxyapatite.

Small cubes of this material can be grafted into the damaged living bone site. Over a period of time, new bone develops and grows in and around the implant. Eventually, successful repair of the damaged bone is achieved. Although not yet tested on humans, the indications are that, in the future, this method of bone repair could be used, but only in non-load bearing settings.

Electronic Substrate and Package Ceramics

Electronic substrate and package ceramics are advanced industrial materials that, owing to their insulating qualities, are useful in the production of electronic components.

Modern electronics are based on the integrated circuit, an assembly of millions of interconnected components such as transistors and resistors that are built up on a tiny chip of silicon. In order to maintain their reliability, these circuits depend on insulating materials that can serve as substrates (that is, the bases on which the microscopic electronic components and their connections are built) and packages (that is, the structures that seal a circuit from the environment and make it a single, compact unit). The insulating properties of ceramics are well known, and these properties have found application in advanced ceramic materials for substrates and packages.

Materials

Among the ceramics employed as electronic substrates and packages, the dominant material is alumina (aluminum oxide, Al_2O_3). The advantages of alumina include high resistivity, good mechanical and dielectric strength, excellent thermal and corrosion stability, and the ability to provide hermetic seals. Its major disadvantages are a relatively high dielectric constant (which delays signal propagation) and low thermal conductivity (which makes it inefficient at drawing away heat). For these reasons ceramic materials with improved properties are under development.

Substrates

Alumina substrates are made as either thick films or thin films. Thick-film substrates are made by tape-casting processes (primarily doctor blading) to thicknesses of less than 1.5 millimetres, or they are made by dry pressing at thicknesses greater than 1.5 millimetres. The formulation contains 96 percent Al_2O_3, the rest being alkaline-earth silicates that form a glass during sintering to bond the alumina particles together. The glass also serves to bond with thick-film resistors or conductors built on the surface of the substrate.

Thin-film substrates (less than 1 millimetre thick) are also made by doctor blading. Organic binder and plasticizer are added to assist in forming. The standard formulation is 99.5 percent alumina, with small amounts of magnesia (MgO) and silica (SiO_2) added to control grain size during sintering.

Multilayer Packages

Integrated circuits are often contained in multilayer packages such as chip carriers, dual-in-line packages, and pin-grid arrays. These structures serve to house semiconductor devices in strong, thermally stable, hermetically sealed environments.

Ceramic packages are made of 90–94 percent Al_2O_3, the rest of the formulation consisting of glass-forming alkaline-earth silicates. One major requirement is that the formulations be able to be cofired with tungsten or molybdenum metallization lines. The alumina layers are produced by tape casting/doctor blading, after which the tapes can be hole-punched or laser-cut, via-hole-coated (vias are conductive pathways between layers), and metallized with tungsten or molybdenum by screen printing. Several layers are then laminated into multilayer structures. Cofiring takes place at temperatures up to 1,600 °C (2,900 °F) in protective atmospheres of hydrogen or hydrogen-nitrogen gas in order to prevent the metals from oxidizing. The result of cofiring is a monolithic package with internal conductor paths. The silicon chip is mounted in the package, and the package is hermetically sealed with a glass or metal lid.

The purpose of the integrated circuit package is to contain the silicon device and to connect it to the external electric circuitry. The packaging materials must have low dielectric constants (in order to minimize the delay in signal processing), and they must conduct heat away from the semiconductor devices. Alumina is poor on both counts. Higher-thermal-conductivity materials exist, but they are either toxic (as in the case of beryllium oxide, BeO) or are poor cofiring ceramics (*e.g.*, aluminum nitride, AlN). Glass-ceramic formations have been developed that are easy to process, have low dielectric constants, and also match the thermal expansion coefficients of high-conductivity metals (gold and copper) that are used in electric circuitry. However, they have low strengths and low thermal conductivities.

Other Types of Ceramics

Ceramics is a category of hard material that is typically manufactured by heating minerals. Humans have produced ceramics since at least 24,000 BC. This predates the use of metal. Modern ceramics include some of the strongest known materials. Ceramics are commonly used in construction, consumer products, and vehicles, scientific and industrial. The following are common types of ceramics.

Bone China

A type of porcelain made with bone ash as a key ingredient. This is a British innovation that emerged at the 18th century. For about 150 years, bone china was produced almost exclusively produced in the UK with Jap the market in the mid-20th century. Ceramics is often defined as an inorganic material but bone china is an exception.

- Properties: Bone china has similar properties to other porcelains but is known for its superior chip resistance, whiteness and translucency.
- Uses: Bone china is considered a premium material due to its history of being manufactured in the UK to produce high quality tableware and ornamental items.

Earthenware

The earliest form of human produced ceramics that was created in fire pits as early as 29,000 BC. Earthenware with selected clays and other materials and is not heated to the point of vitrification that gives porcelain and ceramics their glassy, translucent and nonporous properties. As such, earthenware is opaque, porous and is softer than porcelain.

- Properties: Low mechanical strength. Easily chipped and scratched. Requires less energy to produce than most other ceramics and is relatively inexpensive. An opaque color that is often similar to clay with browns, reds and oranges being common. Earthenware must be glazed to be watertight.
- Uses: Inexpensive tiles and pottery such as flower pots.

Stoneware

Stoneware is fired at a temperature above earthenware but below porcelain. This results in a vitreous or semi ceramic that is watertight but not as translucent and white as porcelain.

- Properties: Stoneware is often valued for its earthy look and relative strength as compared to earthenware. Its properties are similar to porcelain beyond its appearance. As such, it has similar uses including in industrial equipment where it is used for its resistance to chemicals.
- Uses: Stoneware is used to make both expensive and inexpensive pottery and tiles.

Fired Bricks

Bricks produced by heating minerals such as sand and clay are ceramics.

- Properties: The properties of fired bricks differ widely based on their composition and manufacture. Generally speaking, they are durable, heavy, brittle and can withstand high temperatures.
- Uses: Walls, landscaping, fireplaces and chimneys.

Carbon

A common definition of a ceramic is a hard material that is held together with ionic and covalent bonds. According definition, elemental carbon is a ceramic. Additionally, carbon based materials such as carbon fiber, carbon nanotubes and graphene can be considered ceramics. Graphene is currently considered the strongest known material.

- Properties: Carbon materials such as carbon fiber and graphene have a high specific surface area meaning that they can be formed into sheets that are extremely light weight. In theory, you can make a 2630 m^2 sheet of graphene that only weighs one gram (28309 square feet with a weight of 0.035274 ounces). These materials have extremely high tensile and compressive strength and display remarkable electron mobility.
- Uses: Advanced composite materials such as carbon reinforced plastic are used for high performance applications such as aviation, sporting goods and safety equipment. Carbon fiber also has broad applications in areas such as textiles, technology products and construction.

Silicon

As with carbon, elemental silicon can be considered a ceramic due to its chemical properties. Silicate minerals extremely abundant as they make up approximately 90 percent of the Earth's crust. Clays and sand that are used to creat common ceramics are often silicon based. For example, the kaolinite used to make porcelain and the silica used to make fired bricks is silicate minerals.

- Properties: A hard and brittle crystalline solid and semiconductor.
- Uses: High purity crystalline silicon such as polycrystalline silicon is used to make solar panels, integrated circuits and other semiconductor devices. Silicon minerals are used to make glass, ceramics and are used as an aggregate in cement. As such, they are amongst the most common raw materials used in construction.

Silicon Carbide

A semiconductor material containing silicon and carbon that occurs naturally as the very rare mineral moissanite.

- Properties: A very hard and durable ceramic and semiconductor that occurs in about 250 crystalline forms. Naturally colorless but often colored by impurities such as iron. Exhibits low thermal expansion.
- Uses: Furnaces, cutting tools, abrasives, brake disks, electrical power systems, heating elements and lighting. The natural form of silicon carbide, moissanite, is valued as a jewel as it has a similar hardness and appearance to diamonds. Synthetic moissanite is used a diamond substitute that is generally harder than cubic zirconia.

Titanium Carbide

An extremely hard and heat resistant ceramic that has a black color.

- Properties: Hard, heat resistance and resistance to wear and corrosion.
- Uses: Machine parts, tool bits, watch mechanisms and heat shields.

Tungsten Carbide

A material made with equal parts of tungsten and carbon that is dense and hard.

- Properties: Hard, dense, strong with low electrical resistivity.
- Uses: Cutting tools, industrial equipment and sports equipment.

Barium Titanate

A ceramic valued for its electrical properties.

- Properties: Barium titanate is a strong dielectric, an electrical insulator that can be polarized by an applying an electric field. It is also a piezoelectric material, meaning that it accumulates an electrical charge in response to mechanical stress.
- Uses: Capacitors, transducers, microphones and other sensors.

Boron Carbide

A hard ceramic that is resistant to heat, ionizing radiation and chemicals.

- Properties: Extremely hard and heat resistant, semiconductor, absorbs radiation and non-reactive to most chemicals.
- Uses: Locks, nozzles, scratch resistant coatings, tools and dies, abrasives, neutron absorbers and brake linings. Often used in composite materials for its hardness.

Ceramic Foam

Foam is any material that contains pockets of gas or liquid. Ceramic foams are typically designed to be strong lightweight or to have special properties for applications such as thermal insulation.

- Properties: Ceramic foams are generally less strong than a solid ceramic but may be very strong relative to their weight. They may be as much as 96% gas by volume. Some ceramic foam is less brittle than their solid counterpart because air pockets may prevent cracks in the material from spreading.
- Uses: Lightweight structural material, thermal insulation, acoustic insulation, filtration and absorption of environmental pollutants.

Ferrite

Ferrite is a ceramic material manufactured by heating rust (iron oxide) with one or more metals such as barium nickel and zinc.

- Properties: Ferrites are permanent magnets valued for their magnetic field that attracts or repels other magnetic materials such as iron and other magnets. Ferrites are also poor conductors of electricity.
- Uses: Refrigerator magnets, speakers, electric motors, inductors, transformers, magnetic recording tapes, memory hardware and industrial applications.

References

- Traditional-ceramics, technology: britannica.com, Retrieved 21 May, 2019
- Whiteware, art: britannica.com, Retrieved 8 January, 2019
- Glass-ceramics, sciences, examples, dissertation: ukessays.com, Retrieved 13 May, 2019
- Ceramic-cements: ceramicindustry.com, Retrieved 25 February, 2019
- Structural-ceramics, materials-science: sciencedirect.com, Retrieved 16 January, 2019
- Electroceramics, technology: britannica.com, Retrieved 29 March, 2019
- Piezoelectric-ceramics, materials-science: sciencedirect.com, Retrieved 30 April, 2019
- Electronic-substrate-ceramics, technology: britannica.com, Retrieved 29 June, 2019

3

Ceramic Engineering

The branch of engineering which is involved in creating objects from non-metallic and inorganic materials is known as ceramic engineering. Some of the applications of ceramics are ceramic inks, ceramic tiles and ceramic capacitors. The diverse aspects of ceramic engineering along with these applications have been thoroughly discussed in this chapter.

Ceramic engineering is the technology that involves the design and manufacture of ceramic products. Ceramics are inorganic, nonmetallic materials that have been hardened by baking at high temperatures. Highly regarded for being resistant to heat, these materials can be used for many demanding tasks that other materials, such as metals and polymers, cannot.

Until about the mid-twentieth century, the most important ceramics were the traditional clays, which were made into pottery, dinnerware, bricks, tiles, and decorative items. Since then, new materials called advanced ceramics have been prepared and are being used for a wide range of applications, including components used by the aerospace, automotive, defense, environmental, fiber-optic, and medical technologies. Ceramic parts are also used in cellular phones and personal computers. In addition, ceramic engineering is useful in such areas as petroleum refining, food processing, packaging, and mining.

The space shuttle has a coat of ceramic tiles that protect it from the searing heat produced during reentry into the atmosphere.

Traditional ceramic raw materials include clay minerals such as kaolinite. Modern ceramic materials include silicon carbide and tungsten carbide, both of which are highly resistant to abrasion and are used in applications such as the wear plates of crushing equipment in mining operations. Each of NASA's Space Shuttles has a coating of ceramic tiles that protect it from the searing heat (up to 2,300 °F) produced during reentry into Earth's atmosphere. Thus, ceramic engineering is an important contributor to the modern technological revolution.

A ceramic material may be generally defined as any inorganic crystalline oxide material. It is solid and inert. Ceramic materials are brittle, hard, strong in compression, weak in shearing and tension. They withstand chemical erosion that occur in an acidic or caustic environment. In many cases withstanding erosion from the acid and bases applied to it. Ceramics generally can withstand very high temperatures such as temperatures that range from 1,000 °C to 1,600 °C (1,800 °F to 3,000 °F). Exceptions include inorganic materials that do not have oxygen such silicon carbide. Glass by definition is not a ceramic because it is an amorphous solid (non-crystalline). However, glass involves several steps of the ceramic process and its mechanical properties behave similarly to ceramic materials.

Ceramic Process

The ceramic process generally follows this flow:

Milling → Batching → Mixing → Forming → Drying → Firing → Assembly

1. Milling is the process by which materials are reduced from a larger size to a smaller size. Milling may involve breaking up cemented material, thus the individual particle retain their shape or pulverization which involves grinding the particles themselves to a smaller size. Pulverization is actually fracturing the grains and breaking them down.

Generally, milling is done through mechanical means. The means include attrition which is particle to particle collision that results in agglomerate break up or particle shearing. Compression which is applying compressive forces that result in break up or fracturing. Another means is impact which involves a milling media—or the particles themselves—that cause break up or fracturing.

Examples of equipment that achieve attrition milling is a planetary mill or an wet attrition mill, also called wet scrubber. A wet scrubber is a machine that has paddles in water turning in opposite direction causing two vortexes turning into each other. The materials in the vortex collide and break up. Equipment that achieves compression milling include a jaw crusher, roller crusher, and cone crushers.

Finally, impact mills may include a ball mill with media that tumble and fracture material. Shaft impactors cause particle to particle attrition and compression which achieve size reduction.

2. Batching is the process of weighing the oxides according to recipes, and preparing them for mixing and drying.

3. Mixing occurs after batching and involves a variety of equipment such as dry mixing ribbon mixers (a type of cement mixer), Mueller mixers, and pug mills. Wet mixing generally involve the same equipment.

4. Forming is making the mixed material into shapes, ranging from toilet bowls to spark plug insulators. Forming can involve: Extrusion, such as extruding "slugs" to make bricks, Pressing to make shaped parts, or slip casting, as in making toilet bowls, wash basins and ornamentals like

ceramic statues. Forming produces a "green" part, ready for drying. Green parts are soft, pliable, and over time will lose shape. Handling the green product will change its shape. For example, a green brick can be "squeezed," and after squeezing it will stay that way.

5. Drying is removing the water or binder from the formed material. Spray drying is widely used to prepare powder for pressing operations. Other dryers are tunnel dryers and periodic dryers. Controlled heat is applied in this two-stage process. First, heat removes water. This step needs careful control, as rapid heating causes cracks and surface defects. The dried part is smaller than the green part, and is brittle, necessitating careful handling, since a small impact will cause crumbling and breaking.

6. Firing is where the dried parts pass through a controlled heating process, and the oxides are chemically changed to cause sintering and bonding. The fired part will be smaller than the dried part.

7. Assembly, this process is for parts that require additional subassembly parts. In the case of a spark plug, the electrode is put into the insulator. This step does not apply to all ceramic products.

Applications

Ceramics can be used for many technological applications. One example is NASA's Space Shuttle, which uses ceramic tiles to protect it from the searing heat of reentry into Earth's atmosphere. Future supersonic space planes may likewise be fitted with such tiles. Ceramics are also used widely in electronics and optics. In addition to the applications listed here, ceramics are used as a coating in various engineering cases. For example, a ceramic bearing coating may be used over a titanium frame for an airplane. Recently, the field has come to include the studies of single crystals or glass fibers, in addition to traditional polycrystalline materials, and the applications of these have been overlapping and changing rapidly.

Aerospace

- Engines: Shielding a hot running airplane engine from damaging other components.
- Airframes: Used as a high-stress, high-temp and lightweight bearing and structural component.
- Missile nose-cones: Shielding the missile internals from heat.
- Space Shuttle tiles.
- Rocket Nozzles: Withstands and focuses the exhaust of the rocket booster.

Biomedical

- Artificial bone: Dentistry applications, teeth.
- Biodegradable splints: Reinforcing bones recovering from osteoporosis.
- Implant material.

Electronics and Electrical Industry

- Capacitors,
- Integrated Circuit packages,
- Transducers,
- Insulators.

Optical

- Optical fibers; Glass fibers for super-fast data transmission,
- Switches,
- Laser amplifiers,
- Lenses.

Modern-day Ceramic Engineering

Modern-day ceramic engineers may find themselves in a variety of industries, including mining and mineral processing, pharmaceuticals, foods, and chemical operations.

Now a multi-billion dollar a year industry, ceramics engineering and research has established itself as an important field of science. Applications continue to expand as researchers develop new kinds of ceramics to serve different purposes. An incredible number of ceramics engineering products have made their way into modern life.

Ceramic Technology

Ceramic Technology is an emerging branch of Engineering/Technology dealing with the study of the properties, manufacture, design and applications of ceramic materials. The ceramic designers are concerned with designing various types of ceramic wares used in industries, households, institutions, corporate houses and other places.

The use of ceramic ware can be for both decorative as well as functional purposes. Ceramic products include table and kitchenware, decorative ware like flower vase, insulators, potters, building materials like tiles, etc. Cement is also considered as one of the ceramic materials.

A ceramic technologist is concerned with study, research and development of the ceramic materials, planning and developing processing systems, designing and construction of plants and machinery along with other engineers, planning and implementation of processing methodologies, designs and development of various applications like glass, porcelain, cement, refractors, insulators, composite materials, enamels, cement etc.

Industrial Ceramics

Industrial ceramics covers fields of industrial development such as improved processing cycles and advances in design and production in machinery, equipment and systems for the fabrication of all kinds of ceramics, from clay products to space age ceramic materials. It is needed to serve as a concrete link between the achievements of applied research and current production practice. Particular emphasis is placed on those aspects of design and plant technology which are directed towards obtaining unit cost reductions through better use of material, energy and labor resources, and assure high standards of product quality and reliability.

Ceramics exhibit useful properties as high strength and hardness, high melting temperatures, chemical inertness, and low thermal and electrical conductivity but that also display brittleness and sensitivity to flaws. As practical materials, they have a history almost as old as the human race. Traditional ceramic products, made from common, naturally occurring minerals such as clay and sand, have long been the object of the potter, the brick maker, and the glazier. Modern advanced ceramics, on the other hand, are often produced under exacting conditions in the laboratory and call into play the skills of the chemist, the physicist, and the engineer. Containing a variety of ingredients and manipulated by a variety of processing techniques, ceramics are made into a wide range of industrial products, from common floor tile to nuclear fuel pellets. Yet all these disparate products owe their utility to a set of properties that are universally recognized as ceramic-like, and these properties in turn owe their existence to chemical bonds and atomic structures that are peculiar to the material. The composition, structure, and properties of industrial ceramics, their processing into both traditional and advanced materials, and the products made from those materials are the subject of many articles on particular traditional or advanced ceramic products, such as whitewares, abrasives, conductive ceramics, and bioceramics.

Tribology of Ceramics

Characteristics of friction and wear of a ceramic material are determined by a combination of its bulk microstructure parameters, surface conditions and environmental factors (temperature, atmosphere pressure, etc.), lubrication conditions.

- Effect of microstructure on tribological properties of ceramics:
  - Parameters of microstructure and their influence on friction and wear of ceramics.
  - Manufacturing processes forming microstructure of ceramics.
- Effect of surface characteristics on tribological properties of ceramics:
  - Surface characteristics.
  - Methods of modification of ceramic surfaces.
- Effect of lubrication on tribological properties of ceramics.

Effect of Microstructure on Tribological Properties of Ceramics

Grain Size

The main disadvantage of Ceramics as compared to Metals and Polymers is their low fracture toughness. Toughness is a bulk mechanical property of a material however it correlates with its wear resistance particularly when the wear is a result of abrasive action caused by cracking. Finer grain structure results in increased toughness and better wear resistance. Grain size also determines the surface finish quality, which may be achieved by grinding and polishing operations. Fine grain structure allows to decrease the size of the surface micro asperities after the surface finish operation resulting in lower coefficient of friction.

Critical Flaw Size

The size of a flaw that results in rapid fracture. Effect of flaw size on the fracture strength of a ceramic material is expressed by the Griffith equation:

$$\sigma_C = K_{IC} / (Y(\pi a)^{1/2})$$

where,

- K_{IC} – stress-intensity factor, measured in MPa*m½;
- a – the flaw size;
- Y – geometry factor.

According to the equation flaws of lower size result in increased material toughness and higher wear resistance. Flaw size is generally proportional to the grain size.

Homogeneity

Homogeneous distribution of the matrix particles size and pores size, second phase particles (toughening particles) incorporated between the matrix particles, aid phase (binders) locating at the grains boundaries results in lowering the flaw size and consequently in increase of the fracture strength (according to the Griffith equation). Higher fracture strength causes higher wear resistance.

Bulk homogeneity of the microstructure allows creating fine and homogeneous surface finish with low content of surface flaws. High quality surface possess low coefficient of friction.

Manufacturing Processes Forming Microstructure of Ceramics

- Powder preparation: Powder characteristics such as particle shape (spherical, irregular), average particle size, and size distribution determine the ceramic grain size and the amount and size of the pores.
- Compaction (shape forming): The value of the applied pressure, the method of its application (Uniaxial (Die) Pressing, Isostatic Pressing, Injection Molding, Extrusion, Slip Casting, etc.) and the amount of binders and other additives (plasticizers, lubricants, deflocculants, water etc.) determine the pores size and the residual internal stresses.
- Sintering: Diffusion proceeding during sintering process causes the pores to diminish or even to close up resulting in densification of the ceramic material. The bonding and other second phases are distributed between the grains of the main ceramic phase. The matrix grains may grow during the sintering process. Thus sintering process determines the final grains and pores size and the physical and chemical homogeneity.

Effect of Surface Characteristics on Tribological Properties of Ceramics

1. Surface topography: Friction characteristics (coefficient of friction, wear) are strongly dependent on the type of the lubrication regime (boundary lubrication, mixed lubrication, hydrodynamic lubrication). The lubrication regime is determined by the ratio of the lubricant film thickness to the surface roughness Ra.

Rough ceramic surface with relatively large micro asperities causes direct contact between the rubbing surfaces and results in high coefficient of friction and increased wear. High surface finish quality allows improving the tribological characteristics of ceramics. Ceramics are brittle and they wear by fracture mechanism, which is characterized by formation of cracks in the subsurface regions surrounding the wear groove. The volume of the lost material is higher than the volume of the wear track. Thus wear of brittle ceramics results in roughening the surface. The effect of roughening during friction is lower in toughened ceramics.

2. Surface defects: Sintering defects, surface machining, impacts during friction, embedded particles introduce surface flaws, which lead to fracture cracking and increase wear.

3. Surface composition and tribochemical reactions: Ceramic surface may adsorb molecules of the environmental gases and liquids. Such surfaces with modified composition may have different coefficient of friction. Coefficient of friction of ceramics in vacuum is commonly higher than that in air.

Hydration of Oxide ceramics in a humid atmosphere also results in changing their coefficients of friction and wear. Wear of hydrated silicon nitride and silicon carbide is decreased. Wear of hydrated Alumina ceramics and Zirconia ceramics is increased due to chemisorption embrittlement.

Surface of Non-oxide ceramics oxidizes in the presence of Oxygen in the environment. The oxidation is enhanced at increased temperatures. Oxide film on the surface of a non-oxide ceramic decreases the coefficients of friction serving as a solid lubricant.

Methods of Modification of Ceramic Surfaces

- Plasma oxidizing: A method of surface oxidation by elemental Oxygen supplied to the ceramic surface by plasma.
- Ion nitriding and carburizing: A method of introducing nitrogen (nitriding) or carbon (carburizing) atoms into the ceramic surface by means of plasma (glow-discharge).
- Ion implantation: A method of introducing a material into a ceramic surface by electrostatically accelerated ions.
- Laser densification: A method of heating the ceramic surface layer by a laser beam resulting in closing the pores between the ceramic powder particles.
- Electron beam densification: A method of heating the ceramic surface layer by an electron beam resulting in closing the pores between the ceramic powder particles.
- Chemical etching: Cleaning the ceramic surface by acids.
- Sputter etching: Bombarding the ceramic surface by accelerated plasma ions, which vaporize the surface molecules.

Effect of Lubrication on Tribological Properties of Ceramics

Lubricants decrease coefficient of friction and reduce wear of the rubbing parts. Lubricants remove the heat generated by friction. This function is particularly important for ceramics since they have lower thermal conductivity and usually produce more heat due to relatively high coefficient of friction. Lubricants remove wear debris from the rubbing surfaces. Lubricants also protect the ceramic surface from the environment.

1. Liquid lubricants: Liquid hydrocarbon lubricants are commonly used for relatively low temperatures (up to 392 °F/200 °C). Silicone oils may be used up to 570 °F (300 °C).

2. Solid lubricants: Solid lubricants may be used for lubricating ceramics in various forms: suspensions in liquid lubricants, dry powders, Dispersions in gases, coatings. Requirements to solid lubricants properties: good adhesion to the ceramic surface, low shear strength in the sliding direction and high compression strength in the direction of the load (perpendicular to the sliding direction).

Substances used as solid lubricants: graphite, molybdenum disulfide, boron nitride, Polytetrafluoroethylene (PTFE), calcium fluoride-barium fluoride eutectic. Maximum work temperature some of the solid lubricants is low (PTFE: 392 °F/200 °C). Other lubricants may withstand up to 1508 °F/820 °C (calcium fluoride-barium fluoride eutectic).

3. Gaseous lubricants: Vapors of some organic substances may serve as lubricants for ceramics. The vaporized molecules of such lubricant reach the ceramic surface react with it and form on its surface a film possessing low coefficient of friction.

Ceramic-matrix Composites

The class of materials known as ceramic matrix composites, or CMCs, shows considerable promise for providing fracture-toughness values similar to those for metals such as cast iron. Two kinds of damage-tolerant ceramic-ceramic composites are being developed. One incorporates a continuous reinforcing phase, such as a fiber; the other, a discontinuous reinforcement, such as whiskers. The major difference between the two is in their failure behavior. Continuous-fiber-reinforced materials do not fail catastrophically. After matrix failure, the fiber can still support a load. A fibrous failure is similar to that which occurs in wood.

Incorporating whiskers into a ceramic matrix improves resistance to crack growth, making the composite less sensitive to flaws. These materials are commonly described as being flaw tolerant. However, once a crack begins to propagate, failure is catastrophic.

Of particular importance to the technology of toughened ceramics has been the development of high-temperature silicon carbide reinforcements.

SiC fibers, which are capable of withstanding temperatures to about 1,200 °C, are manufactured from a polymer precursor. The polymer is spun into a fine thread, then pyrolized to form a 15- ∝ m ceramic fiber consisting of fine SiC crystallites and an amorphous phase. An advantage of the process is that it uses technology developed for commercial fiber products such as nylon and polyester. Two commercial SiC fiber products are Ube Industries' Tyranno fiber and Nippon Carbon's Nicalon fiber, both from Japan.

SiC filaments are prepared by chemical vapor deposition. A thick layer of silicon carbide is deposited on a thin fiber substrate of tungsten or carbon. Diameter of the final product is about 140 ∝ m.

Although developed initially to reinforce aluminum and titanium matrices, SiC filaments have since been used as reinforcement in silicon nitride. The material is manufactured by Avco Specialty Materials/Textron in the U.S. and by Sigma Composite Materials in the Federal Republic of Germany.

SiC whiskers consist of a fine (0.5 ∝ m-diameter) single-crystal structure in lengths to 100 ∝ m. The material is strong (to 15.9 GPa) and is stable at temperatures to 1,800 °C. Whiskers can be produced by heating SiO_2 and carbon sources with a metal catalyst in the proper environments. These reinforcements are manufactured on a commercial scale in Japan by Tateho Chemical Industries and Tokai Carbon Co.

Although these materials are relatively new, at least one successful commercial product is already being marketed. A SiC-whisker-reinforced alumina cutting-tool material is being used to machine nickel-based superalloys. In addition, considerable interest has been generated in reinforcing other matrices such as mullite, silicon carbide, and silicon nitride for possible applications in automotive and aerospace industries.

Interface Conditions

In addition to developments in reinforcement materials, advances in controlling the interfacial bond between matrix and reinforcement have led to further mechanical property improvements of ceramic-ceramic composites. The interfacial bond must be optimized to promote favorable toughening mechanisms such as crack deflection and crack bridging. Without proper interface control, a brittle polyphase material results, rather than a toughened composite.

Toughness improvements by interfacial modifications have been made in both fiber and whisker-reinforced systems. Interface control has resulted in the development of toughened fiber-reinforced glass-ceramic matrix materials at the United Technologies Research Center and of toughened fiber-reinforced zirconia-based materials at the Naval Research Laboratories. At present, interfacial control is more advanced for fiber-reinforced composites than for whisker-reinforced materials.

One current approach is to design the interface so it has a parting layer that promotes crack deflection parallel to the fiber length. The parting layer protects the fiber from damage by deflecting cracks, enabling the undamaged reinforcement to support load and bridge cracks during matrix failure. Thus, the composite does not fail catastrophically. Fracture morphology is comparable to that of the fibrous fracture of wood structures. Current materials being used for such interfaces are boron nitride and carbon, materials that have weak crystallographic orientations that preferentially delaminate.

Modifications to the interfacial zone of whisker-reinforced composites are in their developmental infancy because of the difficulty of applying thin coatings on fine whiskers. Studies at Oak Ridge National Laboratories have demonstrated that thermal treatments of whiskers prior to their incorporation into an alumina matrix can increase fracture toughness of the composite. In those materials, best toughness - about 8.0 MPavm - results with whisker surfaces modified to be carbon rich and oxygen poor.

Some investigators feel that the approaches used by the carbon-carbon community, such as applying various CVD coatings to seal off the fibers, may result in near-term solutions for improving toughness of fiber-reinforced ceramics.

In whisker-reinforced materials, the matrix usually seals off the interface region from the composite exterior. This protects the interface from oxidizing environments. However, once cracks are initiated, they allow access of atmospheric elements into the interior. As with fiber-reinforced materials, new interface compositions must be developed that are stable in oxidizing environments.

In addition, there is still a need to develop further understanding of the role of whisker interfaces on toughening mechanisms for ceramics. The requirements of fiber and whisker-reinforced systems appear to have many similarities.

Reinforcement Needs

Although the current interest in ceramic-matrix composites has resulted from improved reinforcements, there is still a need for further developments. Specifically, reinforcements are needed for ceramic matrices for service at temperatures greater than 1,800 °C.

Currently available polymer-derived fibers are limited because they deteriorate above 1,200 °C. A program aimed at developing higher temperature fiber has been sponsored by the Department. of Defense, combining the expertise of Dow Corning in silicon-based materials with that of Celanese in fiber technology. From this program has come a new fiber material that has higher thermal stability than commercially available fibers.

SiC filament material has limitations in oxidizing environments due to its carbon core and carbon surface coatings that oxidize above 600 °C. These filaments are designed for use in aluminum and titanium matrices. A similar product, engineered for ceramic matrices, is needed.

SiC whiskers are nearer-term reinforcement for commercial ceramic-matrix composites, having already demonstrated success in reinforcing alumina. As with the other reinforcing materials, the whiskers currently being produced are more appropriate for reinforcing metals. Current theory indicates that thicker whiskers (1 to 3 ∝ m) are more appropriate for ceramics. Such materials are now under development.

Dimox Process

Ceramic matrix composites are steadily moving from the laboratory to initial commercial applications. For example, engineers are currently evaluating these materials for use in the hot gas zones of gas turbine engines, because ceramics are known for their strength and favorable creep behavior at high temperatures. Advanced ceramics, for example, can potentially be used at temperatures 400 to 900 °F above the maximum operating temperature for superalloys.

Until recently, however, there has been more evaluation than implementation of advanced ceramics for various reasons. Monolithic or single-component ceramics, for example, lack the required damage tolerance and toughness. Engine designers are put off by ceramic material's potential for catastrophic, brittle failures. While many CMCs have greater toughness, they are also difficult to process by traditional methods, and may not have the needed long-term high-temperature resistance.

A relatively new method for producing CMCs developed by Lanxide Corp., Newark, Delaware, promises to overcome the limitations of other ceramic technologies. Called the Dimox directed metal oxidation process, it is based on the reaction of a molten metal with an oxidant, usually a gas, to form the ceramic matrix. Unlike the sintering process, in which ceramic powders and fillers are consolidated under heat, directed metal oxidation grows the ceramic matrix material around the reinforcements.

Examples of ceramic matrices that can be produced by the Dimox directed metal oxidation process include Al_2O_3, $Al_2Ti)_5$, AlN, TiN, ZrN, TiC, and ZrC. Filler materials can be anything chemically compatible with the ceramic, parent metal, and growth atmosphere.

The first step in the process involves making a shaped preform of the filler material. Preforms consisting of particles are fabricated with traditional ceramic forming techniques, while fiber preforms are made by weaving, braiding, or laying up woven cloth. Next, the preform is put in contact with the parent metal alloy. A gas-permeable growth barrier is applied to the surfaces of this assembly to limit its shape and size.

The assembly, supported in a suitable refractory container, is then heated in a furnace. For aluminum systems, temperatures typically range from 1,650 to 2,100 °F. The parent metal reacts with the surrounding gas atmosphere to grow the ceramic reaction product through and around the filler to form a CMC.

Capillary action within the growing ceramic matrix continues to supply molten alloy to the growth front. There, the reaction continues until the growing matrix reaches the barrier. At this point, growth stops, and the part is cooled to ambient temperature. To recover the part, the growth barrier and any residual parent metal are removed. However, some of the parent metal (5 to 15% by volume) remains within the final composite in micron-sized interconnected channels.

Traditional ceramic processes use sintering or hot pressing to make a solid CMC out of ceramic powders and filler. Part size and shapes are limited by equipment size and the shrinkage that occurs during densification of the powders can make sintering unfeasible. Larger parts pose the biggest shrinkage problem. Advantages of the directed metal oxidation process include no shrinkage since matrix formation occurs by a growth process. As a result, tolerance control and large part fabrication can be easier with directed metal oxidation.

In addition, the growth process forms a matrix whose grain boundaries are free of impurities or sintering aids. Traditional methods often incorporate these additives, which reduce high-temperature properties. And cost comparisons show the newer process is a promising replacement for traditional methods.

Self-lubricating Ceramic Composites

Strong ionic or covalent bonds between the atoms of ceramic materials determine their strength, hardness, high melting point, modulus of elasticity (rigidity), temperature and chemical stability.

In many friction applications where high strength, corrosion resistance and refractoriness should be combined with good tribological properties (coefficient of friction, wear resistance. Ceramic materials have an excellent resistance to different types of wear due to their high strength. Coefficient of friction is lower in high surface finish ceramics with fine grain structure.

Liquid lubrication allows dramatically decrease the coefficient of friction. Solid lubricants continuously supplied to the rubbing ceramic surface have a similar effect. However use of lubricants is impossible in the applications in which the ceramic part operates under the conditions of high temperature, vacuum or corrosive environment.

Continuous supply of a solid lubricant to the ceramic surface may be provided by the ceramic material itself if it is a composite material with the ceramic (e.g., alumina) matrix containing dispersed particles of the lubricant (self-lubrication).

During the friction action the solid lubricant smears over the contact surface and forms a transfer film reducing the coefficient of friction.

Solid lubricants, which are used or may be potentially used in self-lubricating ceramic composites:

1. Graphite;
2. Hexagonal boron nitride;
3. Sulphides, selenides and tellurides (chalcogenides) of metals (e.g., Molybdenum disulfide);

4. Oxides: B_2O_3, MoO_2, ZnO, Re_2O_7, TiO_2, CuO;
5. Soft metals (bismuth, tin, silver, indium, lead).

The examples of self-lubricating ceramic composites:

1. Alumina-graphite composite: Graphite is a solid lubricant relating to the class of Inorganic lubricants with lamellar structure. The layered structure allows sliding movement of the parallel planes. Alumina-graphite composite may be prepared by Isostatic Pressing. The fine grain alumina-20%graphite composite has the coefficient of friction 0.25, which is a half of that of a pure alumina ceramic.

2. Silicon nitride-graphite composite: Coefficient of friction and wear rate of silicon nitride-graphite composite was researched by A. Gangopathyay, S. Jahanmir and M.B. Peterson ("Self-lubricating ceramic matrix composites", Friction and wear of ceramics). It was observed that the coefficient of friction of silicon nitride was reduced from 0.43 to 0.2 with the addition of graphite. The wear rate was not changed with the addition of graphite up to 36%. Further increase of the graphite content led to increase of the wear rate.

3. Alumina-carbon nanotube composite: Carbon nanotubes (CNTs) are carbon allotropes having cylindrical (tube-like) crystal structure. Hot pressed alumina-CNT composites have higher fracture toughness than monolithic alumina ceramic due to the ability of the nanotubes to absorb energy through their highly flexible elastic behavior during deformation.

Tribological effect of carbon nanotubes on the alumina matrix composites was studied by An, You and Lim, ("Tribological properties of hot-pressed alumina CNT composites", Wear of materials). Increase of CNT content from 0 to 12% caused reduction of the coefficient of friction from 0.5 to 0.3. The lubricating effect of CNT in the composite is explained by the rolling motion of the nanotubes over the contacting surface.

4. Alumina-CaF_2 composite: Calcium difluoride (CaF_2) (fluorite, fluorospar) possesses lubrication properties and can be used as solid lubricant operating at increased temperatures 750-1750 °F (~400-950 °C). It was found that CaF_2 releases and smears on the contacting surface acting as a solid lubricant. Presence of calcium fluoride in the composite causes decrease of the coefficient of friction. Wear rate of the ceramic composites also decreases with the increase of the CaF_2 content up to 10%.

Ying Jin, Koji Kato and Noritsugu Umehara studied the lubrication effect of CaF_2 in alumina and alumina-silver based composites and found that calcium fluoride causes reduction of the coefficient of friction only in the temperature range 200-650 °C (392-1202 °F).

5. Nano-structured alumina-FeS composite: FeS (Iron (II) sulfide, ferrous sulfide) has the hexagonal crystal structure similar to that of Graphite hexagonal boron nitride and molybdenum disulfide. Materials having hexagonal crystal structure are characterized by low shear strength and therefore low coefficient of friction.

X. Ma and D. Xiao investigated the tribological effect of FeS in the plasma sprayed nano-structured alumina-titania-FeS composite coating. It was found that the coefficient of friction is decreasing with the increase of the FeS content up to 40-50%. The wear rate of the composite containing FeS is lower than that of single FeS and of Al_2O_3-TiO_2 ceramic.

3D Printing Ceramics

It is becoming more practical for potters and ceramic artists or entrepreneurs to take on projects never before possible because of the increasing accessibility of 3D printing of ceramic materials or of materials that can act as piece molds or block molds. Objects themselves can be printed directly by extruding layers of a ceramic paste from a nozzle and by fusing powder particles layer-by-layer. This is an additive process as opposed to subtractive (where material is cut away from a block to create a 3D object). The latter is more practical for making molds of relief designs for pressing the faces of tiles or for ramp pressing plastic clay.

The practicalities of additive processes and quality and shapes that can be made are still big limiting factors to the technology. However it is clear that refractories (like shelves, posts, supports and even entire kilns) can also be printed and the smoothness of finish is not nearly as important. By printing a honeycomb structure within the refractory they can be made very light and well insulating and can be made from much more expensive materials than would otherwise be possible. Even things like stainless steel can be printed, this enables making complex molds for use with plastic and dust pressing processes.

Many technologies must be understood and exploited to make this possible in an application. One of the most difficult to surmount is learning the software design and conversion tools. This can be very confusing since hundreds of products are available. However there are standards and printers expect to receive a specific file format: STL. 3D design software can be very expensive but there are open source solutions. A major enabler has been a policy change by Auto Desk, the maker of industry standard tools for many years. They are building their future of mechanical design around Fusion 360 and, as of 2015, are releasing it for free use by education and business earning less than 100,000 per year. This product has exceptional online resources and training and this development could be the most important single factor the puts 3D into prime-time ceramic production.

Another factor is the RepRap international movement to develop open source hardware and software platforms for 3D printing. Reprap printers use standard buy-at-a-hardware-store parts or ones that the printer itself can make. This means that anyone can buy and assemble a printer to learn many details of their mechanics and operation. Once understood a printer of any size could theoretically be constructed. In ceramics the focus is on the print head and how to deliver a thick paste to it, extrude it and a constant rate and be able to turn the flow on and off in an instant. The latter can be a real challenge and one solution is to pressure-feed the paste to a print head having a barrel and auger like a pugmill and control the auger using a stepper motor to feed the actual extruder nozzle.

There is a problem with scaling to a bigger size. The printhead or platform must be able to move on at least two axes. There must be a minimum of mass resisting movement in order to have precise and quick movement. But clay is heavy and if the printhead is full of it it cannot be responsive (e.g. Delta printer designs). Likewise if the item being printed is heavy the bed on which it sits cannot move too quick (because the clay, or paste, is soft). And, lower layers of very soft clay must be able to support the entire piece, this obviously limits size. And any degree of departure from vertical side walls.

A major challenge is making the clay set fast enough so that the next layer applied over it will have a firm base. When plastics are extruded in typical home 3D printers they simply cool and firm up, but clay pastes are soft and fragile. Additions of polymers can help set them after extrusion. Augmenting the water with alcohol (e.g. ethanol is preferred over isopropyl alcohol because it carries off more water as it evaporates) speeds evaporation. Pure ethanol and clay is flammable and the plasticity and dry strength are poor. But 50:50 water:ethanol is not flammable and workability is excellent. However, even though the clay feels cold evaporation does not proceed that quickly, it needs a fan to stiffen up fast enough. Under greater pressure stiffer pastes of lower water content can be delivered. When large objects are printed, fans and the extra time between layer deliveries may be enough to enable structural integrity. In some types of 3D printing, support structures of a different material are printed with the item and these are later removed, this could be practical for ceramics also.

An exciting technology is laser fusing of powder, even metal powder. In this way stainless steel can be printed. This enables printing complex molds for dust pressing of tiles.

Notwithstanding all of this, printers designed for clay are appearing on the market (with lots of excitement) and more are coming. You could easily spend $10k but use caution; the laws of physics and common sense apply. Machines have differing priorities. Those that must push clay through a thin tube to a tiny pugmill in the printhead (e.g. the Lutum and WASP machines) will obviously need soft clay and to be reloaded more often and they may not work well with deflocculated bodies (the auger having trouble getting traction). The Potter Bot is moving the entire piece constantly (on the x-y axes), and when that piece gets large it becomes rather like a big cube of jello on a plate being jerked around. Obviously the body will need to be stable or it will just collapse. The only solution is often to simply print really slowly. This can make the novelty of 3D printing clay wear off pretty fast. Many are shocked when they realize that even normal printing time for a large piece could be ten hours.

Don't be stuck with a fancy machine and no clay that works with it. And, you can be sure; the manufacturers are going to follow the revenue model of ink jet printers so brace yourself when you find out the price of cartridges. It might be best to make your own bodies, that are what the pros are doing. You need a propeller mixer to blend the powder and water (more powerful mixers will do this much better). Run the mixer until all air bubbles have surfaced (to de-air it) and all particle surfaces are wetted (this could take 15 minutes). Then pour it on a plaster bat to dewater to the needed consistency. A hand extruder can be used to create the diameter needed to feed the machine (the clay needs to be soft). The character and suitability of the body will be a big part of any success you have, and understanding the recipe and being able to control it will give you a big edge (especially if you want to incorporate alcohol).

Clay suppliers are also producing clays for this. But be skeptical. One of the bodies claims to have 40% 80 mesh grog; however our testing found no grog. Besides, grog is undesirable since it would wear out the print head quickly. Porcelain has 25% water yet claims to have drying shrinkage of 6.5% (this is highly improbable and even if true would make it too non-plastic and fast drying). Some have shipped in, at considerable cost, non-plastic 3D printing clays and found that the plastic bodies they have used for years work better. When printing taller objects it holds up better. When printing take a long time the bottom and top of a piece are soft and the center gets stiffer, plastic clay that dries slower is better. In addition, printing overhangs is a big issue so plastic clays that hold up better are needed.

Because of the difficulty of preparing the clay (because it needs to be so soft, bubble-free and homogeneous) suppliers are going to introduce cartridge solutions. The advantage will be more stable and repeatable performance and a ready-to-use product with good strand adhesion and plastic strength.

3D Printing Polar Ice Porcelain.

To make your own body follow the same pattern. Start with and existing recipe for plastic pottery clay (your body manufacturer may give you the generic recipe of the body you already use). Many are finding that plastic bodies are best; they stick together well, hold up and do not dry too fast. To increase plasticity of a body add bentonite (or vice versa), change kaolins to ball clays or use a more plastic kaolin. Adequate silica is needed so glazes are easy to fit (usually 20% or more). And it will need enough feldspar to make the body vitreous. For example, the popular 50:25:25 recipes for cone 10 are 50 clay, 25 feldspar and 25 silica. For cone 6 it would be closer to 45:35:20. Since it is practical to make your own paste, consider trying the Zero3 porcelain for cone 03 (it contains frit to make it vitrify).

Computer board on a common RepRap 3D printer.

This controls all the stepper motors and the heating element and watches temperature and position sensors. It runs open source software that knows how to interpret an STL file. As it reads that file steps the z-axis upward for each slice and then prints that layer by moving the printhead and movable bed for the x and y axes.

Movable printing bed on a common 3D RepRap printer.

Objects are printed on a platform that moves along the y-axis. The bed is attached to bushings that run along stainless steel rods. Its position is controlled by a rubber belt that feeds around a pulley in the front and around a gear on a stepper motor at the back. It is heated to prevent printed layers from hardening too rapidly or the piece warping during printing.

Printhead of a common RepRap printer.

The assembly consists of stepper motor with its own cooling fan and a heated brass nozzle mounted in a small aluminum block (at the bottom). The nozzle has a heat sensor and its own cooling fan). A plastic filament feeds down through a hole in a laser-cut aluminum spring loaded part. It has an attached roller that forces the filament against a gear fastened to the motor shaft. When the motor steps it pulls in the filament and feeds it down into the heated print head below. The entire head assembly is screwed to a plate that is in turn screwed to bushings that are pulled along the x-axis by a belt controlled by another stepper motor. The computer can thus control the rate of filament feed, the temperature of the nozzle and the x-position of the entire head.

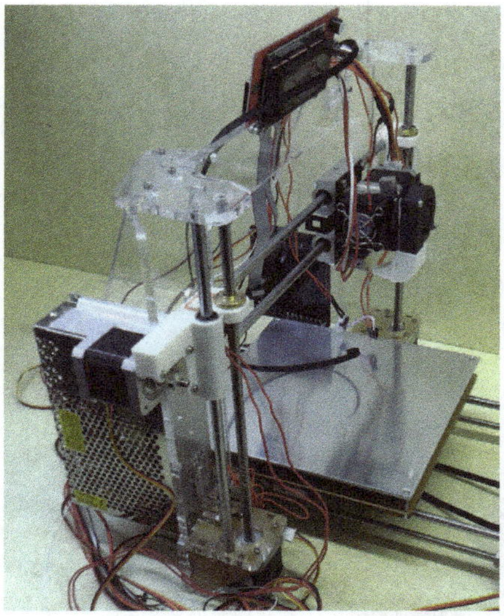

X and Z axis stepper motors on a RepRap printer.

In this printer (which is being assembled) the printhead moves along two stainless steel rods (for the x-axis). Its position is controlled by the front top stepper motor which has a gear through which runs a rubber belt attached to the printhead. The two lower stepper motors with worm gears attached to their shafts control the vertical z-axis position of the printhead assembly. Since the computer controls these motors it can move the head to any position on the x or z axis. Vertical z-movement is slower and more precise since it determines the thickness of each slice to be printed.

3D Printed mug prototype. Although made on an inexpensive Reprap.

An example of how handy the ability to print in 3D can be. The worn-out stainless propeller costs $300 to replace. But the size and pitch of the blades is not right anyway. So we draw them using Fusion 360 and print them in PLA plastic, enabling experimenting with different sizes and pitches. While we could have one printed in stainless at shapeways.com, we do not need to because these plastic ones are surprisingly durable. How about getting a tight fit on the shaft? We measured this shaft with calipers and printed that size. It was a little tight so we printed slightly larger and it fits very tightly. One issue: If you mix slurries with hot water, it will travel up the shaft and the blades will bend.

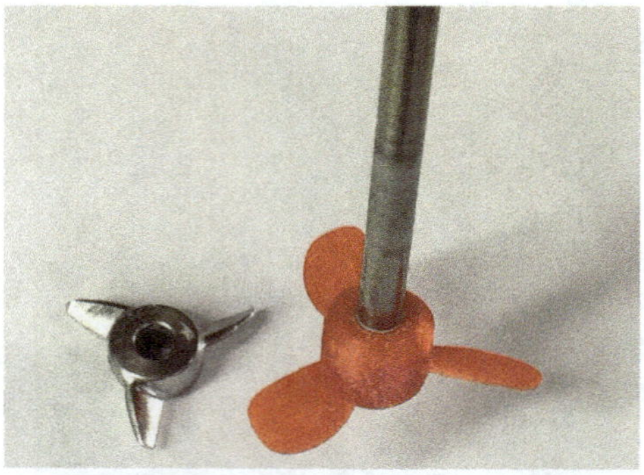

Printing a prototype propeller for my Lightnin lab mixer.

Making ceramic tile shapes by 3D printing your own cookie cutters.

This was done on an affordable RepRap printer. The red plastic templates were drawn in illustrator, extruded in Fusion 360 and sliced and printed using Simplify3D (which took about 30 minutes each). The round wooden block was used to press these cookie-cutters into the clay. The plastic wrap made sticking a non-issue (and rounds the corners nicely). The clay is a low fire, buff burning talc body. Commercial bottled glazes were applied by brushing (in three coats) after bisque. The tiles were fired at cone 03. This is an old classic design that we discovered when researching Damascus tile. The toughest obstacle was learning how to use Fusion 360. It turns out that cookie cutters are a starter project for many 3D software packages, there are lots of videos on making them.

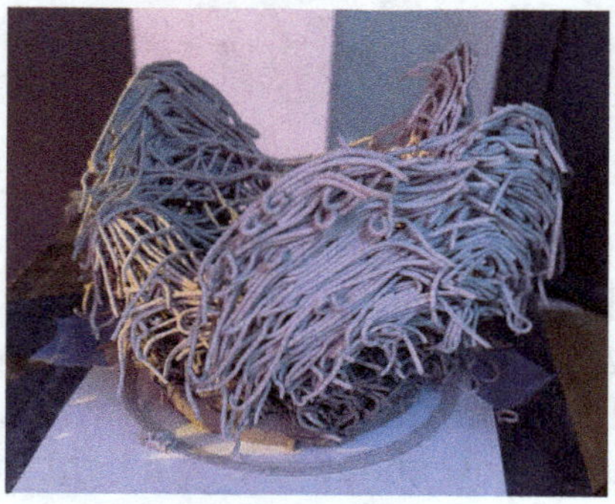

Overnight 3D print goes wrong.

At some point during the night the base could not support the layers being added and it collapsed. The printer happily just kept printing in midair for the rest of the night.

Nanoceramics

Nanoceramics is a term used to refer to ceramic materials fabricated from ultrafine particles, i.e., less than 100 nm in diameter. In this field, a great deal of research has been accomplished in the last 20 years and has resulted in significant outcomes that are of great impact academically as well as industrially.

It has been well proven that the bulk behavior of materials can be dramatically altered when constituted of nanoscale building blocks. Mechanical, magnetic, optical, and other properties of materials have been found to be favorably affected. Hardness and strength, as an example, can be greatly enhanced by consolidating ceramic materials from nanoscale particles. Ductility and superplastic-forming capabilities of nanophase ceramics have now become possible, leading to new processing routes that will be more cost-effective than traditional methods.

Preparation

Remarkable progress in synthetic chemistry has led to significant advances in material science, making possible the synthesis of various substances and materials. The manufacture of ceramics involves heat treatment of tightly squeezed powders. The size of the building block of these powders has been found to affect the properties of the final product. The method of preparation is very often a determining factor in shaping the material and its properties. For example, burning Mg in O_2 (MgO smoke) yields 40-80 nm cubes and hexagonal plates, whereas thermal decomposition of commercial $Mg(OH)_2$, $MgCO_3$, and especially $Mg(NO_3)_2$ yields irregular shapes often exhibiting hexagonal platelets. Surface areas can range from 10 m^2/g (MgO smoke) to 150 m^2/g for $Mg(OH)_2$ thermal decomposition. On the other hand, aerogel-prepared $Mg(OH)_2$ can lead to MgO with surface areas as high as 500 m^2/g.

Physical Methods

Vapor Condensation Methods

Gas-condensation techniques to produce nanoparticles directly from a supersaturated vapor of metals are among the earliest methods for producing nanoparticles. They generally involve two steps: First, a metallic nanophase powder is condensed under inert convection gas after a supersaturated vapor of the metal is obtained inside a chamber. Second, the powder is oxidized by allowing oxygen into the chamber (to produce metal oxide powder). A subsequent annealing process at high temperatures is often required to complete the oxidation. The system consists of a vapor source inside a vacuum chamber containing a mixture of an inert gas, usually argon or helium, mixed with another gas, which is selected based on the material to be prepared. Oxygen is mixed with the inert gas to produce metal oxides. NH_3 is usually used to prepare metal nitrides and an appropriate alkane or alkene, as a source of carbon, is usually used to prepare metal carbides. Nanoparticles are formed when supersaturation is achieved above the vapor source. A collection surface, usually cooled by liquid nitrogen, is placed above the source. The particles are transported to the surface by a convection current or by a combination of a forced gas flow and a convection current, which is set up by the difference in the temperature between the source and the cold surface. Some improved systems involve a way to scrap the nanoparticles from the cold collection surface so that the particles would fall into a die and a unit where they can be consolidated into pellets. Supersaturated vapor can be achieved by many different vaporization methods. The most common techniques include thermal evaporation, sputtering, and laser methods. A variety of nanoscale metal oxides and metal carbides have been prepared using laser-vaporization techniques.

The advantages of vapor condensation methods include versatility, ease in performance and analysis, and high-purity products. On the other hand, they can be employed to produce films and coatings. Furthermore, laser-vaporization techniques allow for the production of high-density, directional, and high-speed vapor of any metal within an extremely short time. Despite the success of these methods, they have the disadvantage that the production cost is still high because of low yields. Heating techniques have other disadvantages that include the possibility of reactions between the metal vapors and the heating source materials.

Spray Pyrolysis

This technique is known by several other names including solution aerosol thermolysis, evaporative decomposition of solutions, plasma vaporization of solutions, and aerosol decomposition. The starting materials in this process are chemical precursors, usually appropriate salts, in solution, sol, or suspension. The process involves the generation of aerosol droplets by nebulizing or "atomization" of the starting solution, sol or suspension. The generated droplets undergo evaporation and solute condensation within the droplet, drying, thermolysis of the precipitate particle at higher temperature to form a microporous particle, and finally, sintering to form a dense particle.

Different techniques for atomization are employed including pressure, two-fluid, electrostatic, and ultrasonic atomizers. These atomizers differ in droplet size (2-15 mm), rate of atomization, and droplet velocity (1-20 m/sec).

These factors affect the heating rate and residence time of the droplet during spray pyrolysis which, in turn, affect some of the particle characteristics including particle size. For a specific atomizer, particle characteristics, including particle size distribution, homogeneity, and phase composition depend on the type of precursor, solution concentration, pH, viscosity, and the surface tension.

Aqueous solutions are usually used because of their low cost, safety, and the availability of a wide range of water-soluble salts. Metal chloride and nitrate salts are commonly used as precursors because of their high solubility. Precursors that have low solubility or those that may induce impurities, such as acetates that lead to carbon in the products, are not preferred.

The advantages of this method include the production of high-purity nanosized particles, homogeneity of the particles as a result of the homogeneity of the original solution, and the fact that each droplet/particle goes through the same reaction conditions. The disadvantages of spray pyrolysis include the need for large amounts of solvents and the difficulty to scale-up the production. The use of large amounts of nonaqueous solvents increases the production expenses because of the high cost of pure solvents and the need for proper disposal.

Thermochemical Decomposition of Metalorganic Precursors

Flame/Thermochemical processes have been widely used to synthesize nanometer-sized particles of ceramic materials. This is another type of gas-condensation technique with the starting material being a liquid chemical precursor. The process is referred to as chemical vapor condensation (CVC). In this process, chemical precursors are vaporized and then oxidized in a combustion process using a fuel-oxidant mixture such as propane-oxygen or methane-air. It combines the rapid thermal decomposition of a precursor-carrier gas stream in a reduced pressure environment with thermophoretically driven deposition of the rapidly condensed product particles on a cold substrate. The flame usually provides a high temperature (1200-3000 K), which promotes rapid gas-phase chemical reactions.

A variety of chemical precursors can be used including metal chlorides, such as $TiCl_4$ to prepare TiO_2 and $SiCl_4$ to prepare SiO_2, metal-alkyl precursors, metal alkoxides, and gaseous metal hydrides, such as silane as a source of silicon to prepare silica. Chlorides have been the most widely used precursors in the industry and the process is sometimes referred to as the "chloride process." The high vapor pressure of chlorides and the fact that they can be safely stored and handled make them excellent potential precursors. The disadvantages of using chloride precursors are the formation of acidic gases and contamination of the products with halide residues. Flame processes are used industrially to produce commercial quantities of ceramic particulates, such as silica and titania. This is because of the low cost of production as compared to all other methods. The disadvantage of flame synthesis is that the control of particle size (both primary particle and aggregates size), morphology, and phase composition is difficult and limited.

Chemical Methods

Sol-gel Technique

The sol-gel process is typically used to prepare nanometer-sized particles of metal oxides. This process is based on the hydrolysis of metal reactive precursors, usually alkoxides in an alcoholic solution, resulting in the corresponding hydroxide. Condensation of the hydroxide by giving off water leads to the formation of a network-like structure. When all hydroxide species are linked, gelation is achieved and a dense porous gel is obtained. The gel is a polymer of a three-dimensional skeleton surrounding interconnected pores. Removal of the solvents and appropriate drying of the gel result in an ultrafine powder of the metal hydroxide. Further heat treatment of the hydroxide leads to the corresponding powder of the metal oxide. As the process starts with a nanosized unit and undergoes reactions on the nanometer scale, it results in nanometer-sized powders. For alkoxides that have low rates of hydrolysis, acid or base catalysts can be used to enhance the process.

When drying is achieved by evaporation under normal conditions, the gel network shrinks as a result of capillary pressure that occurs and the hydroxide product obtained is referred to as xerogel. However, if supercritical drying is applied using a high-pressure autoclave reactor at temperatures higher than the critical temperatures of solvents, less shrinkage of the gel network occurs as there is no capillary pressure and no liquid-vapor interface, which better protects the porous structure. The hydroxide product obtained is referred to as an aerogel. Aerogel powders usually demonstrate higher porosities and larger specific surface areas than analogous xerogel powders.

Sol-gel processes have several advantages over other techniques to synthesize nanopowders of metal oxide ceramics. These include the production of ultrafine porous powders and the homogeneity of the product as a result of homogeneous mixing of the starting materials on the molecular level.

Reverse Microemulsions Method

The reverse micelle approach is one of the recent promising routes to nanocrystalline materials including ceramics. Surfactants dissolved in organic solvents form spheroidal aggregates called reverse (or inverse) micelles. In the presence of water, the polar head groups of the surfactant molecules organize themselves around small water pools 100 A), leading to dispersion of the aqueous phase in the continuous oil phase.

Reverse micelles are used to prepare nanoparticles by using a water solution of reactive precursors that can be converted to insoluble nanoparticles. Nanoparticle synthesis inside the micelles can be achieved by different methods including hydrolysis of reactive precursors, such as alkoxides, and precipitation reactions of metal salts. Solvent removal and subsequent calcination lead to the final product. Several parameters, such as the concentration of the reactive precursor in the micelle and the weight percentage of the aqueous phase in the microemulsion, affect the properties, including particle size, particle-size distribution, agglomerate size, and the phases of the final ceramic powders. There are several advantages to using this method including the ability to prepare very small particles and the ability to control the particle size. Disadvantages include low production yields and the need to use large amounts of liquids.

Precipitation from Solutions

Precipitation is one of the conventional methods to prepare nanoparticles of metal oxide ceramics. This process involves dissolving a salt precursor, usually chloride, oxy-chloride or nitrate, such as $AlCl_3$ to make Al_2O_3, $Y(NO_3)_3$ to make Y_2O_3, and $ZrCl_4$ to make ZrO_2, in water. The corresponding metal hydroxides are usually obtained as precipitates in water by adding a base solution such as sodium hydroxide or ammonium hydroxide solution. The remaining counter-ions are then washed away and the hydroxide is calcined after filtration and washing to obtain the final oxide powder. This method is useful in preparing ceramic composites of different oxides by co-precipitation of the corresponding hydroxides in the same solution. Solution chemistry is also used to prepare non-oxide ceramics or pre-ceramic precursors that can be converted to ceramics upon pyrolysis.

One of the disadvantages of this method is the difficulty in controlling the particle size and size distribution. Very often, fast and uncontrolled precipitation takes place resulting in large particles.

Chemical Synthesis of Pre-ceramic Polymers Coupled with Physical Processing Techniques

This method is based on the use of molecular precursors, which facilitates the synthesis of nanomaterials containing phases of desired compositions. It involves a chemical reaction to prepare an appropriate polymer, which is then converted into ceramic material upon pyrolysis. Using chemical reactions to prepare the pre-ceramic polymer not only allows for control of phase compositions but also overcomes the limitation of low production yields of the physical methods. This method has been very useful in preparing nonoxide ceramics such as silicon carbide and silicon nitride. The conversion of an organometallic precursor into a ceramic depends on different parameters such as the molecular structure of the precursor and the pyrolysis conditions (temperature, duration, and atmosphere). Metal carbides and metal nitrides have been obtained by pyrolysis of polymers containing the appropriate elements such as Si or Al and C or N. These polymers are called pre-ceramic polymers and are prepared from simpler chemical precursors. A considerable amount of free carbon from the thermolysis process is very often a problem. Silicon carbide (SiC) and silicon nitride (Si_3N_4) are the most studied ceramic materials prepared via this route. They are usually synthesized by the pyrolysis of polycarbosilanes and polysilazanes, for which general structural formulas are shown in figure at temperatures between 1000 °C and 1200 °C.

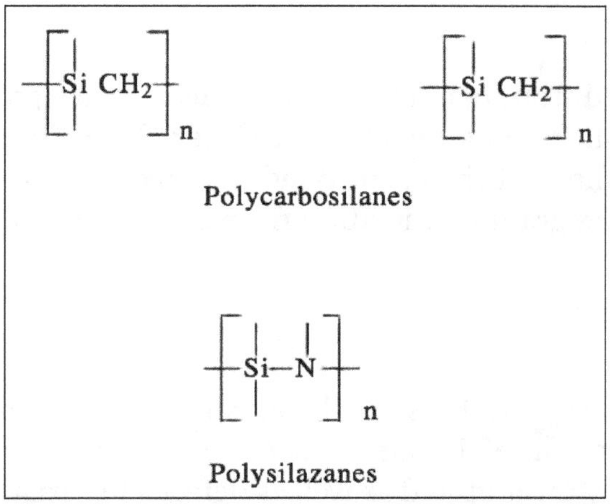

General structural formulas of polycarbosilanes and polysilazanes.

Mechanochemical Synthesis

Mechanochemical synthesis involves mechanical activation of solid-state displacement reactions. This process has been successfully used recently to make nanoceramic powders such as Al_2O_3 and ZrO_2. It involves the milling of precursor powders (usually a salt and a metal oxide) to form a nanoscale composite of the starting materials,which react during milling, and subsequent heating, if necessary, to form a mixture of dispersed nanoparticles of the desired oxide within a soluble salt matrix. Nanopar-ticles of Al_2O_3 (10-20 nm), for example, can be prepared by milling $AlCl_3$ with CaO:

$$2AlCl_3 + 3CaO \rightarrow \gamma\text{-}Al_2O_3 + 3CaCl_2$$

Bonding Considerations

Understanding chemical bonding and structures in ceramic materials is necessary in order to understand their chemical and physical behavior. When materials are composed of nanometer-sized building blocks, they deviate considerably from structural perfection and stoichiometry. As a result, the number of defects due to edges, corners, f-centers, and other surface imperfections is greatly enhanced, which, in turn, affects several physical and chemical properties.

A range of cohesive forces contribute to the nature of bonding in ceramic materials including ionic (MgO, $Fej_{\_x}O$), covalent, metallic, van der Waals, and hydrogen bonding. Ionic compounds are formed when highly electronegative and highly electropositive elements are combined in a lattice. Pure ionic model is a reasonable approximation for some systems while it is a poor approximation for crystals containing large anions and small cations. In such systems, covalent contribution to bonding becomes significant. Van der Waals interactions play a crucial role in many ceramic systems, especially those with layered structures. In many oxide hydrates or hydroxy oxides, hydrogen bonding also contributes to the cohesive energy.

Selected Properties

Ceramics possess their own chemical, physical, mechanical, and magnetic properties that are different from those of other materials such as metals and plastics. The properties of ceramics depend mainly on the type and the amounts of materials in their composition. However, the size of the building blocks of a ceramic material has been found to play an important role in its properties.

When materials are prepared from nanometer-sized particles, a significant portion of the atoms become exposed on the surface. As a result, such materials exhibit unique properties that are remarkably different from those of the corresponding bulk. The physical and chemical properties of nanoparticles show the gradual transition from atomic or molecular to condensed matter systems.

Chemical Properties

Ceramic materials are relatively inert, especially crystalline materials that tend to have perfect structures with minimum amount of defects. Most of the reactivity of these materials involves the surfaces where coordinatively unsaturated as well as defect sites exist. The behavior of the surface toward other species and the nature of interaction depend on the composition and the morphology, which determine the nature and the degree of surface interactions with other substances. Most of the time, interactions are limited to adsorption on the surface, which does not affect the bulk making these materials good corrosion-resistant.

The possibility of preparing ceramic powders in high surface areas with high porosity makes them well desired in some advanced applications. One example is the use of ceramic materials as supports for heterogeneous catalysts. Another example is the use of such materials in biomedical applications, where the surface of nanophase ceramics exhibits a remarkably improved biomedical compatibility compared to conventional ceramics.

Mechanical Properties

Ceramics are very strong materials showing considerable resistance against compression and bending. Some ceramic materials are similar to steel in strength. Most ceramics retain their strength at high temperatures. Silicon carbides and silicon nitrides, as an example, retain their strength at temperatures as high as 1400 °C. As a result, such materials are used in high-temperature applications. Many of the physical and mechanical properties are particle-size dependent. As a result, several systems of nanophase ceramics have exhibited quite interesting and favorably enhanced mechanical properties.

Improved Sintering and Hardness Properties

Nanoceramics are processed from nanophase powders by compacting first powders composed of individual ceramic particles (usually less than 50 nm in size) into a raw shape (often called a green body). This compacted powder is then heated at elevated temperatures. Densification occurs as a result of diffusion of vacancies out of pores (to grain boundaries), which lead to shrinkage of the sample. This process is referred to as pressure-less sintering. Fortunately, nanophase powders were found to compact as easily as their analogous submicron particles. To avoid particle size growth, samples have to be sintered at the lowest temperature possible for a time sufficient to remove the residual porosity and establish coherent grain boundaries. Successful sintering enhances the hardness of the final material.

Experimental evidence shows that nanophase powders densify at faster rates as compared to commercial (submicron) particles. Faster densification rates allow achieving a given density at smaller grain sizes, before serious growth takes place. As a result of their small particle and pore sizes, nanocrystalline powders sinter too much greater densities than their conventional analogs at the same temperature. This also establishes that nanocrys-talline powders, as compared to conventional powders, reach the same density at much lower temperatures. This, of course, eliminates the need for very high temperatures.

One disadvantage that can accompany fast densifica-tion though is inhomogeneous heating where the outside layers of the particles densify into a hard impervious shell which constrains the inside of the sample from normal shrinking, leading to some cracking as a result of strain incompatibility. This problem can be avoided by several ways. The most efficient way is to heat the samples slowly to reduce the shrinkage in the outer shell while heat is transported to the inner regions. On the other hand, high-density nanostructured ceramic systems including Y_2O_3, TiO_2, and ZrO_2 have been achieved by means of pressure-assisted sintering. Applying some pressure during sintering can increase the densification rate and suppress the particle growth.

Nanoscale powders of nonoxide ceramics such as metal carbides and nitrides show similar behavior. Conventional SiC, as an example, is difficult to sinter. Addition of some additives such as boron or carbon is very often necessary to densify SiC. Ultrafine powder of SiC sinters at lower temperatures and densifies without additives. On the other hand, mechanical properties can be fairly improved by the introduction of metallic nanoparticles dispersed within the matrix grains. Such systems are referred to as nanocomposites. Tungsten, nickel, or molybdenum nano-particles dispersed within Al_2O_3 matrix grains, as an example, can enhance the mechanical properties of alumina, including the fracture strength and hardness.

Reduced Brittleness and Enhanced Ductility and Superplasticity

Superplasticity and ductility refer to the capability of some polycrystalline materials to undergo extensive tensile deformation without necking or fracture. Ceramic brittleness is the biggest technical barrier in practical applications, especially in load-bearing applications. Theoretical and experimental results provide evidence for the possibility that traditional brittle materials can be ductilized by reducing their grain sizes. When made from nanoparticles, brittle ceramics can be superplastically deformed at modest temperatures and then heat treated at higher temperatures for high-temperature strengthening.

The capability to synthesize superplastic ceramic materials is now established. Nanocrystalline ceramics deform at faster rates, lower stresses, and lower temperatures. One important use of superplasticity in ceramics is diffusion bonding, where two ceramic parts are pressed together at moderate temperatures and pressures to form a seamless bond through diffusion and grain growth across the interface. Diffusion bonds form more easily in nanocrystalline ceramics than in larger grained ceramics as a result of both the enhanced plastic flow of nanocrys-talline ceramics and the larger number of grain boundaries they provide for diffusional flux across the interface.

Electrical Properties

Ceramics include electrical conducting, insulating, and semiconducting materials. Chromium oxide is an electrical conductor; aluminum oxide is an insulator, while silicon carbide behaves as a semiconductor. As a result, ceramic materials have been used in a variety of electronic applications based on their electrical behavior.

Several electrical properties are particle-size and composition dependent. Electrical resistance and dielectric constant, as an example, for some systems increased as a result of small particle size. Conductivity of some mixed oxide ceramics, such as lithium aluminosilicate, is higher than that of their constituent oxides.

Magnetic Properties

Some ceramic materials possess magnetic properties. These include iron oxide-based ceramics and oxides of chromium, nickel, manganese, and barium. Ceramic magnets are known to exhibit high resistance to demagnetization. As a result, several ceramic powders have been employed in a wide range of electronic and magnetic applications.

The fabrication of such materials from ultrafine particles can significantly enhance their magnetic behavior. The fact that in nanometer-sized particles a large portion of the atoms are on the surface, where the coordination numbers are less than that for bulk atoms, affects several parameters including unique surface/interface behavior and different band structure, which both lead to magnetism enhancement. It is now well established that one of the requirements to achieve appropriate coercivity and high magnetization saturation is to fabricate such materials in highly divided particles, preferably in the nanometer-sized range, with homogeneity and narrow size distribution.

Many other properties are also particle-size dependent. The optical properties, as an example, of some ceramic materials have been found to depend on particle sizes. Nanoparticles of TiO_2, as an example, are more efficient UV absorber than powders of large particles.

Applications

Ceramic materials are of great value in a variety of applications as a result of their unique properties compared to other materials. Because of their electrical and magnetic properties, ceramics are important in several electronic applications, where they are used as insulators, semiconductors, conductors, and magnets. Ceramic materials also have important uses in the aerospace, bio-medical, construction, and nuclear industries. In many of these applications, ceramic materials have shown significantly better performance when fabricated from nanometer- sized particles.

Mechanical Applications

Industrial ceramics are widely employed in applications that require strong, hard, and abrasion-resistant materials. Metal-cutting tools, tipped with alumina, and tools made from silicon nitrides for cutting, shaping, grinding, and sanding iron, nickel-based alloys, and other metals are very commonly used. Other ceramics such as silicon nitrides and carbides are used to make components for high-temperature use such as valves and turbocharger rotors. Ceramic materials and metal-based ceramics (cermets) are used to make components for space vehicles, including heat-shield tiles for the space shuttle and nosecones for rocket payloads.

Electrical Applications

Ceramics are used as insulators, semiconductors, and conductors. Aluminum oxide (Al_2O_3), for example, does not conduct electricity at all and is used to make insulators. Other ceramics, such as barium titanate ($BaTiO_3$), are used as semiconductors in electronic devices. Some copper oxide-based ceramics are superconductive at temperatures higher than those at which metals become superconductive. Superconductivity refers to the ability of a cooled material to conduct an electric current with no resistance. This phenomenon can occur only at extremely low temperatures, which are difficult to maintain. Transition metal nitrides, carbides, and borides are of interest as cathodes in electrochemical applications. This interest stems from the favorable properties of these materials including electronic conductivity and good thermal conductivity, which when coupled with their mechanical strength and high melting points suggest that such materials can be stable in a range of environments.

Table: Examples of electronic ceramics.

Functions	Examples of materials	Applications
Insulation	Al_2O_3, SiC + BeO	IC substrate
Dielectricity	$BaTiO_3$	Capacitor
Semiconducting	SiC, $LaCrO_3$, SnO_2, ZnO + Bi_2O_3	Gas defector, thermistor, varistor
Piezoelectricity	ZnO, SiO_2	Piezolighter, piezofilter, surfacewave transducer,
		Piezovibrator, flexible piezodetector
Pyroelectricity	PZT	IR detectors
Ferroelectricity	PLZT	Optical shutter, optical memory
Ionic conduction	b-Al_2O_3, ZrO_2	Na-S battery, O_2 sensor
Luminescence	Y_2O_2S:Eu, ThO_2:Nd, Al_2O_3:Cr	Cathode luminescence, IR laser
Light guide	SiO_2	Optical communication fiber
Polarization	PLZT	Optical shutter
Soft magnetism	g-Fe_2O_3, $Zn1\_{x}Mn_xFe2O4$	Magnetic tape
Hard magnetism	SrO, $6Fe_2O_3$	Magnet seal

Some ceramics such as strontium titanate ($SrTiO_3$) are employed in the form of thin films as capacitors in several electronic devices because of their capability to store large amounts of electricity in extremely small volumes. Lithium aluminosilicate ceramics have potential applications as solid electrolytes for utilization in solid high-energy density lithium battery systems. Piezoelectric ceramics are now key electronic components for television, FM radio, and the like. Very recently, piezoelectric ceramic displays have been developed in Japan, where microceramic actuators for activating the pixels are used. Piezoelectric effect refers to the appearance of an electric potential across certain faces of a crystal when it is subjected to mechanical pressure.

Other examples of functions and applications of advanced ceramics in the field of electronics are shown in table.

Magnetic Applications

Iron oxide-based ceramics (ferrites) are widely employed as low-cost magnets in electric motors. Such magnets help in converting electric energy into mechanical energy. Unlike metal magnets, ferrites conduct high-frequency currents, and as a result, they do not lose as much power as metal conductors do. Manganese zinc ferrites are used in magnetic recording heads, and ferric oxides are the active component in several magnetic recording media, such as recording tapes and computer diskettes.

Biomedical Applications and Bioceramics

Some advanced materials are used in the biomedical field to make implants for use within the body. The main requirement of biomaterials for this application is the ability of their surfaces to support new bone growth. Ceramic materials are known to possess exceptional biocompatibil-ity properties with bone cells and tissues. Specially made porous ceramic materials such as alumina, titania, zirconia, and others bind with bone and other natural tissues. Such ceramics are used to make hip joints, dental caps, and bridges. Other advanced ceramics such as hydroxyapatite, $Ca_{10}(PO_4)_6(OH)_2$, which is the principal component of bones and teeth, have excellent biocompatibility and bone in-growth capabilities and are used in reconstructing fractured bones and as replacement materials.

Recent research on long-term functions of osteoblasts on nanophase ceramics has shown evidence for unique and significant behavior. Compared to conventional ceramics, nanophase ceramics has shown enhanced osteoblast adhesion and proliferation, alkaline phosphatase synthesis, and concentration of extracellular matrix calcium.

Coatings

Because of its unique hardness and corrosion resistance, ceramic enamel is often used in coating metals. Thin hard ware-resistant coatings of ceramics include materials such as titanium nitride and titanium carbonitride. An emerging class of new hard protecting coatings beyond homogeneous layers of a ceramic nitride material is layered coating structures such as superlattices or multilayers of different nitrides. Such multilayer coating has been successfully applied in several applications such as bearings,pumps, and compressors. Nonnitride coatings, such tungsten carbide/carbon, are also of interest because of their high elasticity and chemical inertness.

Several techniques are being employed for protective ceramic coating. These include thermal spraying, chemical vapor deposition (known as CVD), and plasma spraying.

Nuclear Industry

Lithium-based ceramics are now considered as potential solid tritium breeders in nuclear fusion reactors. Potential breeder materials include $LiAlO_2$, Li_2O, Li_2ZrO_3, and Li_4SiO_4. Solid breeders are safer during operation than liquid lithium systems, which are highly reactive.

A nanoceramic is one of the great outcomes of the evolutionary research in the field of nanoscience and nanotechnology, where fabrication of materials from nanometer-sized building blocks has resulted in a wide range of industrially useful materials. Recent research has proven that ceramic materials fabricated from ultrafine powders can be obtained through several physical as well as chemical methods that can be scaled up to produce commercial amounts. These unique materials have exhibited very remarkable behavior as compared with their bulk counterparts. Significant characteristics include chemical, mechanical, magnetic, electrical, and optical properties. As a result, improved performance of ceramic materials has been observed in a variety of applications including chemical, mechanical, magnetic, electrical, and biomed-ical. The new properties and improved performance of nanoceramics that are being discovered stimulate the development and improvement of ceramic processing, which, in turn, open the doors wide for the use of ceramics in a wide range of new technologies.

Nanocomposite Ceramics

Significant scientific effort has been directed towards making ceramics more flaw-tolerant through design of their microstructures by incorporation of fibers or whiskers which bridge the crack faces just behind the crack tip; by designing microstructures with elongated grains which act as bridges between crack faces just behind the crack tip; by incorporating second phase particles which deflect the crack making it travel a more tortuous path; and by incorporating secondary phases which undergo stress induced volume expansion that forces the crack faces together. However, one of the most recent developments has been the distribution of multiple phases in a ceramic composite at the nanoscopic length scale. Owing to prevalence of nanoscopic features, such composites are referred to as ceramic nanocomposites.

Nanocomposite material has broadened significantly to encompass a large variety of systems such as one-dimensional, two-dimensional, three-dimensional and amorphous materials, made of distinctly dissimilar components and mixed at the nanometer scale. The general class of nanocomposite organic/inorganic materials is a fast growing area of research. Reducing the sizes of structural features in materials leads to a significant increase in the portion of surface/ interface atoms.

The surface/interface energies essentially control the properties of a solid. Interfaces provide a means to introduce non-homogeneity in the material. This non-homogeneity acts as a significant modification of both thermal and mechanical properties of the composites. Selective mixing of materials in a highly tailored morphology with high percentage of interface area, leads to materials with enhanced properties.

The properties of nano-composite materials depend not only on the properties of their individual parents but also on their morphology and interfacial characteristics. The nanocomposites find their use in various applications because of the improvements in the properties over the simpler structures. Few of such advantages can be summarized as:

- Improved mechanical properties e.g. strength, modulus and dimensional stability.
- Decreased permeability to gases, water and hydrocarbons.
- Higher thermal stability and heat distortion temperature.
- Higher flame retardancy and reduced smoke emissions.
- Higher chemical resistance.
- Smoother surface appearance.
- Higher electrical conductivity.

For components used in a gas turbine engine, a lifetime upto 10000 h and a retained strength of ~300 MPa at a temperature of 1400 °C have been postulated, together with negligible creep rate. Furthermore, at elevated temperatures, the material must exhibit high resistance to thermal shock, oxidation, and subcritical crack growth. Ceramic nanocomposites have been shown to be extremely important for such future applications.

Advanced bulk ceramic composite materials that can withstand high temperatures (>1500 °C) without degradation or oxidation can also be used for applications such as structural parts of motor engines, catalytic heat exchangers, nuclear power plants, and combustion systems, besides their use in fossil energy conversion power plants. These hard, high-temperature stable, oxidation-resistant ceramic composites and coatings are also in demand for aircraft and spacecraft applications.

One such material system in this class of composites, Silicon Carbide/Silicon Nitride (SiC/Si_3N_4) composites, have been shown to perform very well under high temperature oxidizing conditions. Interest in such nanocomposites started with experiments of Niihara who reported large improvements in both the fracture toughness and the strength of materials by embedding nanometer range (20-300 nm) particles within a matrix of larger grains and at the grain boundaries. A 200% improvement in both strength and fracture toughness, better retention of strength at high temperatures, and better creep properties were observed.

An advanced nanocomposite microstructure such as that of polycrystalline Silicon Carbide (SiC)-Silicon Nitride (Si_3N_4) nanocomposites, figure contains multiple length scales with grain boundary (GB) thickness of the order of 50 nm, SiC particle sizes of the order of 200-300 nm and Si_3N_4 grain sizes of the order of 0.8 to 1.5 μm. Designing the microstructure of such a composite (and similar others such as TiN-Si_3N_4, SiC-Al_2O_3, SiC-SiC, Graphene/CNT+SiC, and Carbon Fiber+SiC nanocomposites) for a targeted set of material properties is, therefore, a daunting task. Since the microstructure involves multiple length scales, multiscale analyses based material design is an appropriate approach for such a task.

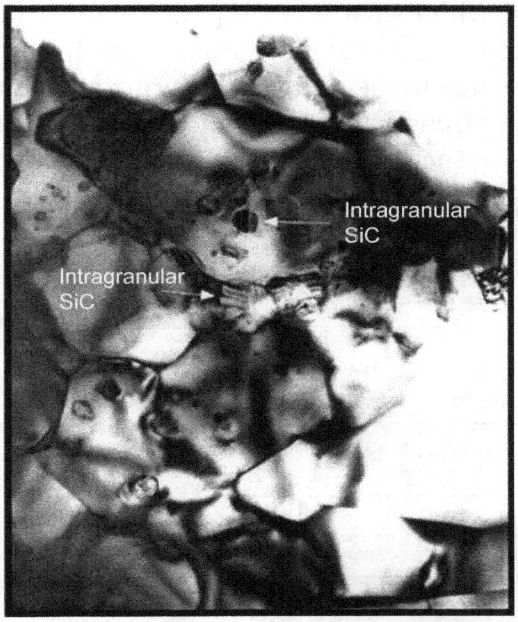

Actual microstructure of a SiC-Si_3N_4 nanocomposite.

The ceramic nanocomposite work in multi-physics Lab at Purdue focuses on: (1) understanding performance of carbide and nitride based high temperature ceramic nanocomposites for extreme environments found in power generation cycles including nuclear applications, (2) multiscale modeling and characterization in oxide ceramic materials and (3) understanding thermal conduction and thermal issues in materials for thermoelectric power generation.

Multiscale Modeling of Ceramic Nanocomposites

The example of work in SiC-Si_3N_4 ceramic nanocomposites are described below:

Multiscale analyses (at nanometer and micrometer length and time scales) based on a combination of CFEM and MD based techniques have revealed that high strength and relatively small sized SiC particles act as stress concentration sites in Si_3N_4 matrix leading to inter-granular Si_3N_4 matrix cracking as a dominant failure mode. CFEM analyses have also revealed that due to a significant number of nano-sized SiC particles being present in micro-sized Si_3N_4 matrix, the SiC particles invariantly fall in wake regions of micro-cracks leading to significant mechanical strength. This finding was confirmed in the MD analyses that revealed that particle clustering along the GBs significantly increases the strength of these nanocomposites. While some nanocomposite morphologies have sharply defined SiC-Si_3N_4 interfaces, other nanocomposite morphologies have diffusion of C, N, or Si atoms at the interfaces.

In the case of SiC-Si_3N_4 nanocomposites, MD analyses have also revealed that the second phase particles act as significant stress raisers in the case of single crystalline Si_3N_4 phase matrix affecting the strength significantly. However, the particle's presence does not have a significant effect on the mechanical strength of bicrystalline or nanocrystalline Si_3N_4 phase matrices. The strength of the SiC-Si_3N_4 nanocomposite structures showed an uncharacteristic correlation between the grain boundary (GB) thickness and temperature.

The strength showed decrease with increase in temperature for structures having thick GBs having diffusion of C, N, or Si atoms. However, for structures with no appreciable GB thickness (no diffusion of C, N or Si atoms), due to the particle clustering and increase in SiC-Si_3N_4 interfacial strength with temperature, the strength improved with increase in temperature. Figure shows the fracture propagation analyses in such nanocomposites obtained using the CFEM.

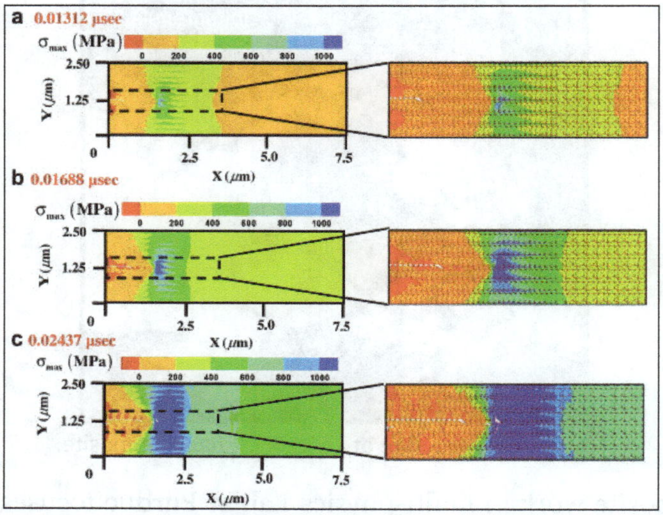

Mesoscale crack propagation and damage propagation in the Si_3N_4 nanocomposites.

The current research work focuses on obtaining experimental images of the ceramic nanocomposites developed by collaborators, developing nanoscale CFEM meshes on such images, and performing failure analyses using the combination of MD and CFEM techniques.

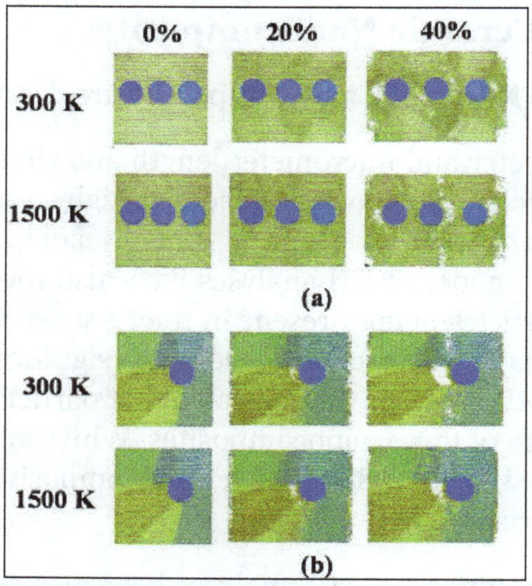

Atomistic damage and failure propagation in two different SiC (particle) and Si_3N_4 (matrix) nanocomposites at two different temperatures.

Petascale Computing based Material Design

Atomistic analyses at the nanoscale can impart important information about the effect of critical features such as a GB, an interface, or a triple junction, etc. on mechanical deformation behavior of a small nanoscale (~ few nm) sample. In multiscale modeling such information is used to formulate macroscale (>few μm) material models for understanding microstructure dependent deformation behavior of a material sample such as the one shown in figure.

Appropriate mathematical models of microstructure property relations allow relating performances like fracture toughness, ultimate strength, fatiguing lifetime, to key material microstructure parameters like volume fraction, particle size, and phase composition. Since a typical nanoscale test sample is much smaller and is subjected to varied surroundings in a typical microstructure, the incorporation of nanoscale information in macroscale models is subjected to statistical uncertainty.

If a complex microstructure is to be designed for a targeted set of properties, it is important that such uncertainties be correctly quantified and incorporated within a robust material design framework.

The figure below details the process flow of a petascale multi-physics model management tool for multiscale material design. Deployed on a petascale machine, the design tool developed in this research, that integrates atomistic and mesoscale analyses using a variable fidelity model management framework, will facilitate a significant reduction in nanomaterials' development cost and time with a simultaneous increase in the possible different combinations of individual composite material phases to achieve desired material performance.

The model management framework, besides managing the models and scales, is also well suited to control hierarchical parallelism. The natural hierarchy is MD within CFEM within design under uncertainty, using a mixed programming model SHMEM by SGI for CFEM and MPI for MD and the uncertainty modeling. Both MD and the uncertainty quantification (via quasi-Monte Carlo integration) can use 1000 processors, and CFEM 10, so 1000 uncertainty quantification groups of 10 CFEM groups of 1000 HMC processors is 10^7 processors, nearing exascale.

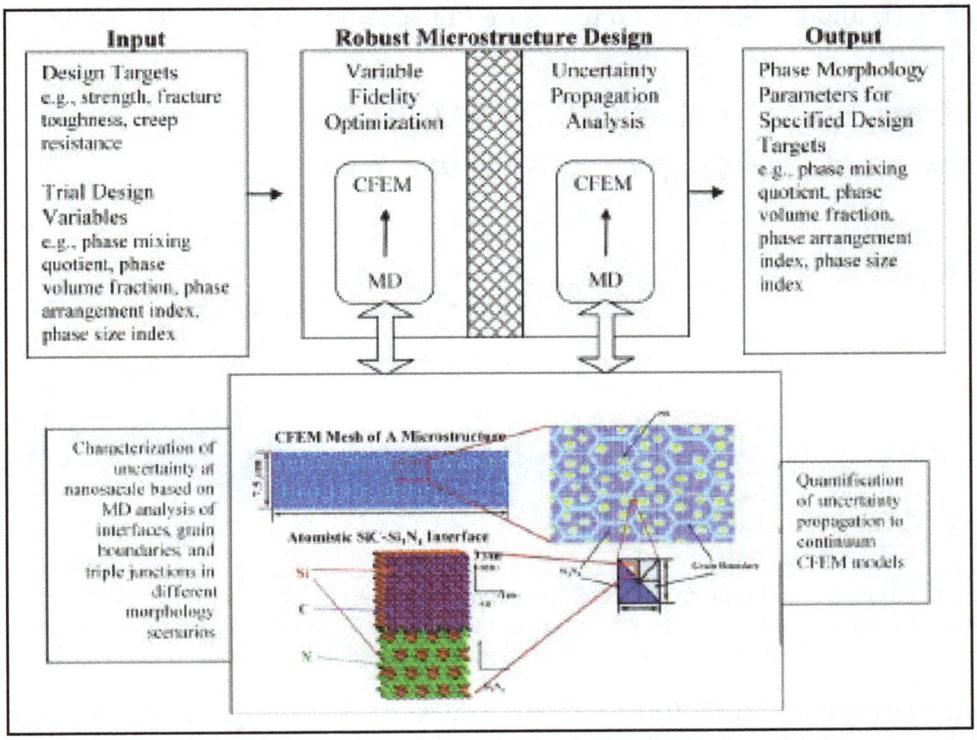

Schematic Petascale Material Design Framework.

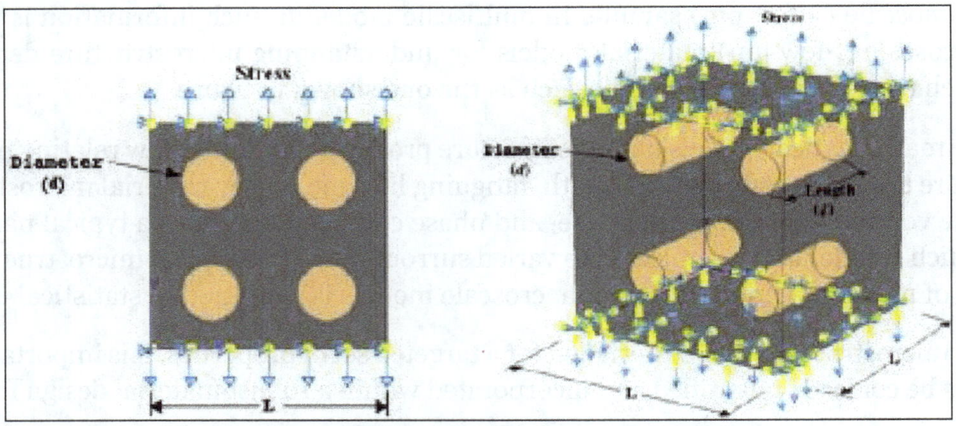

High and low fidelity models for the CFCC nanocomposites.

Preliminary material design analyses of the model system have been performed to understand the morphology related parameters that must be controlled for optimal targeted set of properties. The application of design tool is focusing on the continuous fiber ceramic composite (CFCCs) models of SiC-Si_3N_4 nanocomposites. The second phase (circles and cylinders) are the SiC fibers that have higher elastic modulus and higher creep resistance (E) but lower yield stress and fracture toughness, than that of the primary Si_3N_4 phase. The problem is to design the most suitable CFCC, with maximum strength and creep resistance for a set of external temperatures T, where the number of design variables will depend on whether the simulation tests are run on the 2-dimensional (2-D) or 3-dimensional (3-D) model. The design variables to be considered in the nanocomposite design optimization problem, for the 2-D model, are the fibers diameter (d) and the external temperature (T). And for the 3-D model the design variables to be considered are the fibers diameter (d), the length of fibers (l) and the external temperature (T). The problem definition in standard form is given below:

$$\left.\begin{array}{l}\text{minimize} \quad f(d,l,T)=\{-\sigma_U(d,l,T),\dot{\varepsilon}_0(d,l,T)\} \\ \text{subject to:} \; d_{\min} \le d \le d_{\max} \;\; l_{\min} \le l \le l_{\max} \;\; and \;\; T = 1500\,^{\circ}\text{C}\end{array}\right\}$$

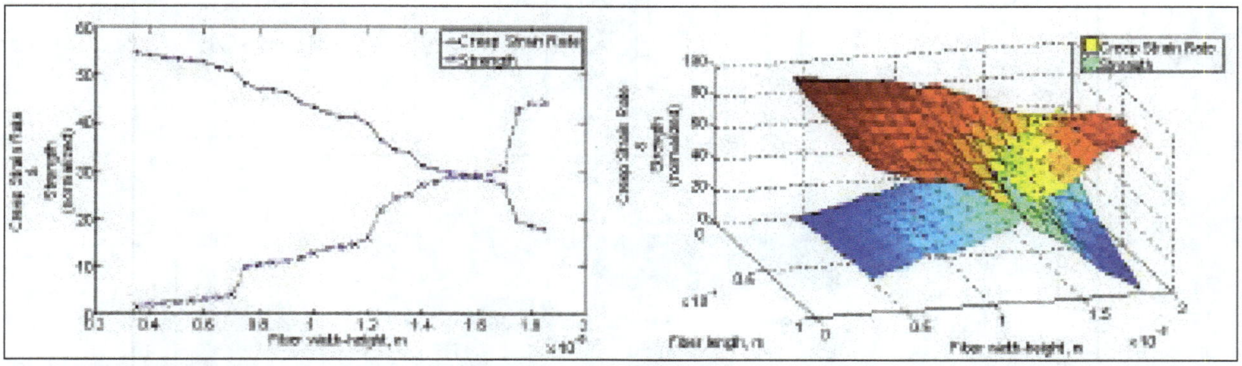

(Left) Strength and creep strain rate at 1500 °C as a function of the design variable width-height (d) for the 2-D low fidelity model. (Right) Strength and creep strain rate at 1500 °C as a function of the design variables width-height (d) and length of fibers (l) for the 3-D high fidelity model.

Figure illustrates normalized (0-100) function values for the strength and creep strain rate as a function of design variables for the high fidelity model (3-D) and low fidelity model (2-D). Figure (left) shows an increase in the CFCC strength and a corresponding decrease in the creep strain rate as the design variable *d* increases. Similarly for the high fidelity model, figure (right) shows an increase in the CFCC strength and a corresponding decrease in the creep strain rate as the design variables *d* and *l* increase.

Fabrication

Focus during this activity is on forming a collaborative modeling-deign-processing framework where complex ceramic nanocomposites for targeted set of mechanical and non-mechanical properties could be produced without wasting significant trial-and-error time and money. Polymer derived ceramics (PDCs) are an attractive approach to make material design predicted morphology of ceramic nanocomposites. First Niihara and his coworkers and then others used this approach to make high performance nanoscale reinforced composites.

Continued research in this area has led to the development of a range of nanostructures. One particularly interesting class of materials has predominantly amorphous Si-O-C nanodomains containing nanoscale SiC and C reinforcements. These materials have the desired characteristics for a broad range of high-temperature applications while offering greater control over processing, compositions and nanostructure. PDCs are produced by pyrolyzing preceramic polymers and are typically amorphous up to very high temperature but provide very intriguing ceramic-like properties, such as good creep and oxidation resistance.

Some of their unique properties are associated with in-situ formation of nanodomains and lack of grain boundaries in their microstructures. Due to the polymeric nature (thermoset) of the precursors, this family of materials is easily processable as fibers, matrices for composites, porous structures and coatings. Most studied PDCs can be categorized into three main groups: (i) silicon carbide (SiC), (ii) silicon oxycarbides (SiOC) and (iii) silicon carbonitrides (SiCN). SiOCs and SiCNs are distinctive due to their hybrid molecular composition between SiO_2 and SiC and between SiC and Si_3N_4, respectively with additional level of "free" carbon as schematically illustrated in Fig for Si-O-C system.

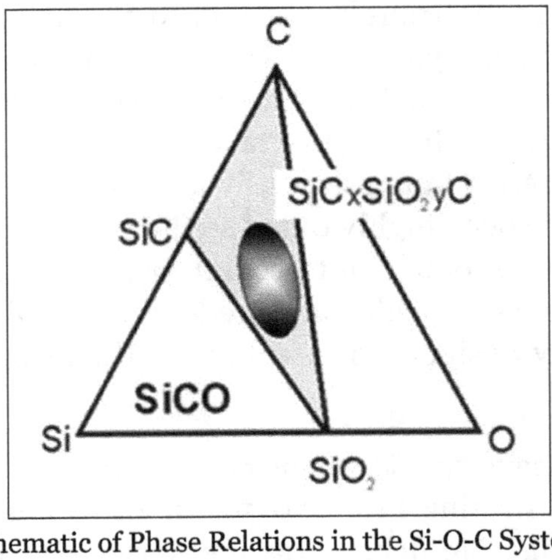

Schematic of Phase Relations in the Si-O-C System.

A unique nanostructural feature of these materials is that the controlled excess carbon is dispersed as graphene layers with domain size of a few nms.

Application of Ceramics

Ceramic Ink

Ceramic inks are simply carriers of ceramic fine particles metallic oxide pigments (not raw colorants but prefired stain powders) that are used in automatic application techniques. Inks must have a physical consistency suitable for producing fine detail, this requires that they be suspended in a medium (as opposed to just water). Depending on the application technique, these mediums can be thick and flow like a printing ink. Or they can have a gel consistency that holds itself in place after application. Or they can be made from nano-size particle stains that stay in suspension in a highly fluid medium (for inkjet printing). They can be water or oil based. They may need to dry quickly, slowly and may not need to dry at all before firing. But the objective is the same: Tune a carrier for the application process so as to achieve a layer of metallic oxide based powder that will produce color exactly where desired and with crisp, well defined color edges.

Ceramic inks also need a melt carrier, that is, the metallic oxide colorant mix must be part of a larger silica:alumina:flux recipe that melts and envelopes it in a glass that will adhere to the body and be compatible with the over-lying glaze (or form a hard ware surface and be compatible with the under-lying glaze). The compatibility must also extend to matching the thermal expansion of the body and glaze and the melt carrier must also have a chemistry that is compatible with the color system. Each different stain system has its own needs and the melt carrier must be tuned to it. Carrier formulations even need to be adjusted with varying proportions of certain colors (to maintain the degree of gloss, for example).

Thus, formulating a good ceramic ink to work with the process at hand can be one of the most challenging tasks a technician will face. The simplest ink system is often used by potters: They mix glycerine with pure ceramic stains for rubber stamping. For some production methods and selected stain systems, this can be successful. The next level of complexity is to blend stains with a glaze base and suspend them in a colorless screen printing medium and then silk screen designs onto tissue paper, and then transfer them onto the ware using water and pressure on the back side. It is now easy to print your own right or wrong-reading negatives on clear emulsion-treated acetates and photo-develop them onto a silk screen. However in recent years, inkjet printing of ceramic stain based inks has all but taken over the entire ceramic world from hobbyist to the largest manufacturers. Many previously unrelated technologies have come together to make this possible. Inks have been highly tuned so they can be used in standard printers. The first major revolution was the ability to print decal transfers directly, this, combined with new software and other developments, brought unprecedented flexibility to the decoration stage of production. But now, specially designed printers spray the design directly on to the finished ware at high speed.

Compared to glazes these inks have very low density. Examples we have noted range from 1.02 to 1.3 (from two suppliers, demonstrating a range of tolerance within the inkjet system). Individual inks must be carefully maintained to density (as measured with hydro gravimeteic density instruments) and viscosity (e.g. Brookfield viscometer). Filtration testing and careful observation of printed tone are also part of the QC regimen.

Ceramic Tile

The tile industry is the largest sector of the ceramic industry. Tile production is done at a wide range of temperatures using all manner of body and glaze types and process methods. While some countries use little ceramic tile, others cover every surface of buildings, even sidewalks with tile. The quality of tile a company produces is a real testament to the expertise of their technical staff; this is because the engineering challenges in tile production are daunting indeed.

Wall and floor tile make up the largest portion of production. Floor tile must be durable, non-slip, dense, strong, easily cleaned, very flat. Wall tile does not have to be any of these, although of course, many types would benefit having one or more of these properties, density especially.

Tile can be formed using many techniques. Hand rolling is surprisingly common in traditional techniques. Tile can be extruded and wet pressed, but by far the most common is dust pressing. Giant hydraulic machines are used to press moist powder (at 500 kg/cm^2) into flats that can be larger than a meter square. Since shrinkage is almost zero no warpage occurs during drying. Glaze can be applied to the dry tile. Firing is done in roller kilns that heat so evenly top and bottom that little or no warping occurs (despite significant firing shrinkage). Not only this, tiles move thru the kiln quickly. This ability to quickly fire such large flat shapes planar is perhaps the greatest achievement in kiln firing technology.

Huge quantities of unglazed porcelain tile are produced. Planarity is a big problem (because porcelain shrinks a lot on firing), so much so that tile have to be ground flat after firing. That alone is an entire industry.

Much tile is covered with a white engobe before glazing. This makes it possible to employ a local clay that fires buff, brown or even red yet have a porcelain-like surface on which to apply glaze. Technicians expend much effort to match firing shrinkage and COE to the body and control slurry properties to get even coverage and good dry bonding.

It would seem that tile needs to be vitreous to be strong and resistant to liquid penetration and crazing but this is not the case. Well fitted engobes and glazes can make it possible to use non-vitrifying bodies of surprisingly high porosity to produce tile of very good quality. Such bodies are easier to fire flat because they shrink less.

The tile industry is at the leading edge of decoration technology. Of course there are many traditional methods of decoration, but today there are just three important words: Ink jet printing. Manufacturers can make a tile look like marble by simply printing a picture of marble on it. Photo realism is possible. Print heads are as wide as the fast moving tiles passing under them and there are separate heads for each color. Entire industries have formed around every aspect of printing (design, ink chemistry, ink rheology, nano particle pigments, machine design).

Technicians fight a constant battle against pinholes. Just one can ruin a tile. Companies must go to incredible lengths in material processing, recipe development and production parameter control to create a pinhole-free tile. They also struggle to keep the tiles flat through drying and firing. Matching glaze and engobe thermal expansion to the body is very important in keeping tiles flat.

Huge supply industries have grown to provide tile manufacturers with everything they need. Cutting edge technology characterizes the equipment industry especially. Suppliers are constantly developing new lines of glazes and novel surface treatments. There are even companies that do nothing but build tile factories, complete and ready to move in and switch on. And, of course, material suppliers (e.g. frits, feldspar, silica, talc, clays, pigments). Suppliers provide enough tech support and expertise that tile manufacturers can even outsource much of their engineering needs.

Ceramic Capacitor

Ceramic capacitors are most commonly found in every electrical device and it uses a ceramic material as the dielectric. The ceramic capacitor is a non-polarity device that means they do not have polarities. So we can connect it in any direction on a circuit board.

For this reason, they are generally much safer than electrolytic capacitors. Here is the symbol for a non-polarized capacitor given below. Many types of capacitors, such as the tantalum bead do not have a polarity.

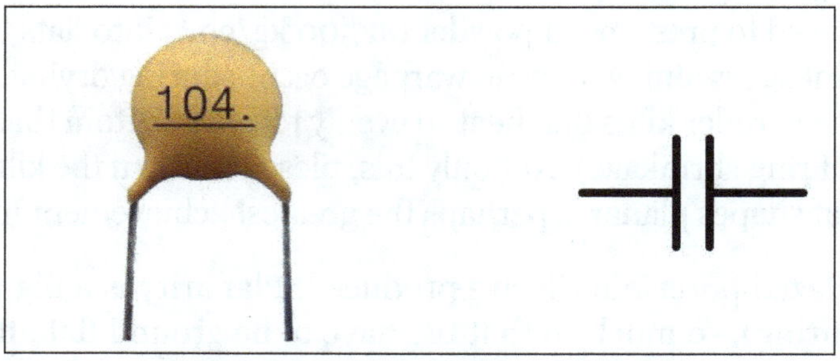

Ceramic Capacitor Polarity and Symbol.

Construction and Properties of Ceramic Capacitors

Ceramic capacitors are available in three types, although other styles are available:

- Leaded disc ceramic capacitors for through hole mounting which is resin coated.
- Surface mounts Multi-layer Ceramic Capacitors (MLCC).
- Special type microwave bare lead-less disc ceramic capacitors that are intended to sit in a slot on the PCB.

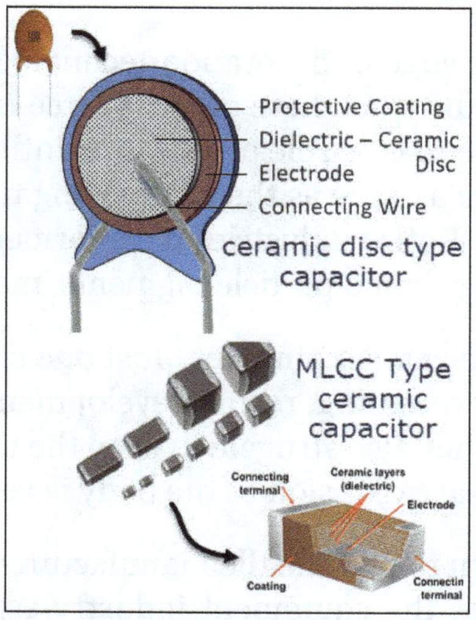

Different types of Ceramic Capacitors.

Ceramic disc capacitors are made by coating a ceramic disc with silver contacts on both sides as shown above illustrates. Ceramic disc capacitors have a capacitance value about10pF to 100μF with a wide variety of voltage ratings, between 16 V to 15 KV and more.

In order to gain higher capacitances, these devices can be made from multiple layers. The MLCCs are made with Paraelectric and Ferroelectric materials mix and alternatively layered with metal contacts.

After completion of the layering process, the device is brought to a high temperature and the mixture is sintered, resulting in a ceramic material of desired properties. Finally, the resulting capacitor basically consists of many smaller capacitors connected in parallel, this leads to increase in capacitance.

The MLCCs consist of more than 500 layers, with the minimum layer thickness of approximately 0.5 microns. As technology progresses, the thickness of the layer decreases and capacitance increases in the same volume.

Ceramic capacitor dielectrics vary from one manufacturer to another, but common compounds include titanium dioxide, Strontium Titanate, and Barium Titanate. Based on the working temperature range, temperature drift, tolerance different ceramic capacitor classes is defined.

Class 1 Ceramic Capacitors

With respect to temperature, these are the most stable capacitors. They have nearly linear characteristics. The most common compounds used as the dielectrics are:

- Magnesium Titanate for a positive temperature coefficient.
- Calcium Titanate for capacitors with a negative temperature coefficient.

Class 2 Ceramic Capacitors

Class 2 capacitors exhibit better performance with respect to volumetric efficiency, but this is at the cost of lower accuracy and stability. As a result, they are normally used for decoupling, coupling and bypass applications where accuracy is not of prime importance.

- Temperature range: -50 °C to +85 °C.
- Dissipation factor: 2.5%.
- Accuracy: Average to poor.

Class 3 Ceramic Capacitors

Class 3 ceramic capacitors offer high volumetric efficiency with poor accuracy and a low dissipation factor. It cannot withstand high voltages. The dielectric used is often Barium Titanate.

- Class 3 capacitor will change its capacitance by -22% to +50%.
- Temperature range of +10 °C to +55 °C.
- Dissipation factor: 3 to 5%.

- It will have a fairly poor accuracy (commonly, 20%, or -20/+80%).

Class 3 types are typically used for decoupling or in other power supply applications where accuracy is not an issue.

Ceramic Disc Capacitor Values

Ceramic disc capacitor code normally consists of a three digit number followed by a letter. It is very easy to decode to find the capacitor value.

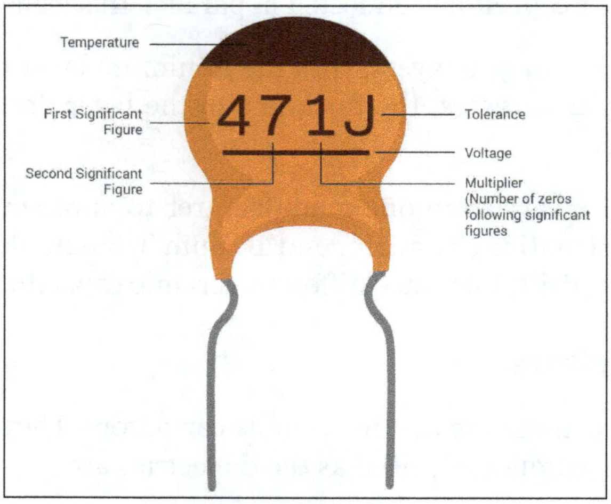

Ceramic Disc Capacitor Values.

The first two significant digits signify the first two digits of the actual capacitance value, which is 47 (the above capacitor).

The third digit is the multiplier (3), which is ×1000. The letter J implies the tolerance of ±5%. Since this is the EIA coding system, the value will be in picofarads. Therefore, the value of the capacitor above is 47000 pF ±5%.

Table: EIA coding system.

1st Digit	2nd Digit		3rd Digit (Multiplier)		Tolerance Letter
0	0	0	×1	C	±0.25 pF
1	1	1	×10	D	±0.5 pF
2	2	2	×100	E	±1 pF
3	3	3	×1000	G	±2%
4	4	4	×10000	J	±5%
5	5	5	×100000	K	±10%
6	6	6	×1000000	L	±15%
7	7	7	×10000000	M	±20%
8	8	8	×100000000	N	±30%
9	9	9	×1000000000	Z	+80% to 20%

For example, if a capacitor is marked as 484N, its value is 480000 pF ±30%.

Applications of Ceramic Capacitors

- Ceramic capacitors are majorly used in the resonant circuit in transmitter stations.
- Class 2 high-power capacitors are used in high voltage laser power supplies, power circuit breakers, induction furnaces etc.
- Surface mount capacitors are often used in printed circuit boards and high-density applications.
- Ceramic capacitors can also be used as a general purpose capacitor, because of their non-polarity and are available in a large variety of capacitances, voltage ratings, and sizes.
- Ceramic disc capacitors are used across brush DC motors to minimize RF noise.
- MLCC used in printed circuit boards (PCB) are rated for voltages from only a few volts up to several hundreds of volts, depending on the application.

Ceramics in Combustion Engines

Ceramics are used or may be potentially used in combustion engines due to their following advantages over Metals:

- High strength;
- Refractoriness (high maximum service temperature). Ceramics retain their properties (including strength and hardness) at elevated temperatures;
- Low density;
- High wear resistance;
- Low coefficient of friction.

Theoretical aspects of Application of Ceramics in Combustion Engines

Thermodynamic Aspect

Maximum efficiency of a heat engine is:

$$\eta_{max} = 1 - T_c/T_h$$

where,

- T_c - temperature of the cold sink (cooling media), K;
- T_h - temperature of the hot source (combustion chamber), K.

According to the expression higher combustion temperature results in better theoretical efficiency of the engine. In addition to this higher temperature causes the fuel burning to be more complete, which also increases the engine efficiency. Metal combustion chamber (cylinder) is capable to operate at maximum 1000 K (1341 °F). If the chamber is made of a ceramic the operation temperature may be increased to 3000 K (4941 °F). Higher temperature in the combustion chamber improves the engine efficiency.

Dynamic Aspect

All combustion engines have accelerating parts (reciprocating and rotating). A certain part of the combustion energy is consumed for driving these parts. The inertia forces developed by the moving parts are proportional to their weights. Therefore decrease of the parts weights will increase the engine efficiency. Ceramics (as compared to metals) have lower density and higher specific strength. Replacement of the materials of the moving parts from metals to ceramics will result in higher engine efficiency.

Tribological Aspect

Ceramics with fine grain structure may be polished to very high surface quality, which provides low coefficient of friction. The main disadvantage of ceramics is low fracture toughness causing increased wear rate by the fracture mode of the abrasive wear when the material cracks in the subsurface regions surrounding the wear groove. Toughened ceramics with homogeneous fine grain microstructure have low wear rate. Depending on the operating temperature liquid lubricants, Solid lubricants or metallic overlays may be used for the lubrication of engine ceramic parts. Current and potential applications of ceramics in combustion engines

Current Application of ceramics in combustion engines:

- Reciprocating engines:
  - Valve guides,
  - Cam follower rollers,
  - Thermal barrier coatings (e.g., exhaust pipes),
  - Turbocharger rotors,
  - Ball bearings,
  - Pump seals,
  - Spark plug insulators.
- Turbine engines:
  - Nozzles,
  - Ceramic lining of combustors,
  - Turbine blades.

Potential applications of ceramics in combustion engines:

- Reciprocating engines:

- Ceramic liners of combustion chambers,
- Pistons,
- Piston rings,
- Cylinder heads,
- Valves and valve guides.

- Wankel engines:
    - Three-sides rotors,
    - Housing (combustion chamber),
    - Side plates.
- Turbine engines:
    - Rotors,
    - Exhaust components,
    - Vanes,
    - Shrouds.

Ceramics in the Aerospace Industry

Ceramic materials are generally lighter than metals, and this low mass makes them highly appealing to the aerospace industry. However, the cost of working with an advanced ceramic material is such that a clear advantage must be established by using it. Once a benefit has been identified for a product or system (e.g., being able to run at a higher temperature or increased electrical activity), a range of ceramics is available.

Electrical Applications

Advanced ceramics underpin the electronics industry, and the average aircraft is packed full of electronics. Gradually, these electrical components, such as sensors, antennas, capacitors and resistors, are getting increasingly smaller and more capable. Therefore, this is a major area of development for advanced ceramics.

As far back as the 1990s, the design team of Concorde, the world's only supersonic airliner at the time, selected a machinable glass ceramic because it needed a lightweight and electrically insulating technical material for use in the engine control and management system. The glass ceramic material is stable at high temperatures and can be machined like plastic, making it an attractive option for this application.

Structural Applications

Structural ceramics (crystalline inorganic non-metals) are used in aerospace as thermal barrier coatings (TBCs) in the hot part of the engine. In addition, these materials are being used in composites either as reinforcement and as a matrix such as in ceramic matrix composites (CMCs). Being lightweight and tough tends to be a main driver for using a ceramic composite. From here, engineers need to assess how a composite will perform at an elevated temperature in an air atmosphere and what impact erosion will have on the system and at what rate.

Ceramics are lighter than most metals and stable at temperatures substantially above high-grade technical plastics. As a result of these and other properties, structural ceramic applications include thermal protection systems in rocket exhaust cones, insulating tiles for the space shuttle, missile nose cones, and engine components. The U.S. Space Shuttle Orbiter program team decided to use the machinable glass ceramic at all hinge points, windows, and doors on the reusable Space Shuttle Orbiter. In addition, large pieces of the glass ceramic have been used in a NASA space-borne gamma radiation detector.

Turbine Applications

Technical ceramics have been used for various parts of the engine for the past 30-40 years, but a lot of activity currently surrounds the development of silicon carbide (SiC/SiC composites) for use in jet engine turbines, mainly concentrated on the turbine blades. The main driver is fuel efficiency, as engineers seek to run the jet engine without the need for cooling channels that currently stop the metal alloy blades from melting. If the blades were made of ceramic composites, which could deal with temperatures of 1,500-1,600 °C, the engine could run at higher temperatures. Energy efficiency would therefore increase, which leads to less fuel and the airplane's ability to fly further or more efficiently.

Huge Potential

Ceramics are integral to the aeronautics and aerospace industries. They are ubiquitous in the electrical systems and facilitate the drive toward more powerful, yet smaller electrical devices. Structural ceramics are increasing in popularity and deployment, and they offer huge potential for transforming aircraft engine capabilities that could dramatically influence the aeronautics of the future.

Ceramics in Dentistry

Bones and teeth, the hard tissues in the human body have an inorganic component. The inorganic component primarily consists of hydroxyapatite ($Ca_{10}(PO_4)_6(OH)_2$, HA). In most cases, they also an organic component which is most often collagen. Enamel, the outer layer of teeth is the hardest material in the body, and thus it is not surprising that it consists of approximately 92% hydroxyapatite.

Requirements of Dental Materials

While having to be stable in the environment described above and able to withstand the loads associated with chewing, dental materials need to satisfy another criteria, aesthetics. With society becoming increasingly self-conscious, any dental materials that are used in visible locations must have color and translucency as close to natural teeth as possible.

Fillings

Traditionally filling has been made from silver/tin/mercury amalgams. However, resin based filling materials are rapidly gaining in popularity for health and aesthetic reasons. These resins are usually filled with 35-85% ceramic fillers such as silicate glasses, colloidal silica or quartz. Despite the advantages of using ceramic filled resins, these materials may be susceptibe to wear when applied to chewing surfaces.

For this reason, all-ceramic inlays, onlays and crowns are also gaining in popularity. CAD-CAM systems are also becoming more popular allow dentists to machine dental ceramic blanks to suit individual patients and fir them in the same visit. Materials that are suitable for this application include:

- Leucite reinforced feldspathic porcelain.
- Alumina with continuously interconnected porosity, infiltrated with lanthanum aluminosilicate glass, after machining to provide translucency.
- Glass infiltrated porous spinel.
- Glass infiltrated porous zirconia.

Crowns and Veneers

Porcelain fused to metal (PFM) materials constitute about 75% all crowns in a market consuming about 30 million crowns in the USA alone. The porcelain used is feldspathic porcelain. The porcelain contains varying amounts of crystallised leucite. The amount of crystallised leucite influences properties such as strength and thermal expansion. Thermal expansion is critical in the manufacture PFM implants to avoid cracking of the ceramic, which could lead to failure during manufacture or in operation.

PFM materials are also used for veneers to cover damaged front teeth and crowns, although ceramic cores are being used in some crowns. All ceramic crowns are normally coated with porcelain so that color and translucency can be matched.

Dental Implants

Dental implants are used as an alternative to bridges where a tooth has been lost or removed. These are also made of PFM, where a biocompatible metal post made out a material such as titanium is anchored into the jaw bone and the porcelain crown is affixed to the post.

The metal post may also be coated with hydroxyapatite to aid bone bonding and rapid osseointegration. Bio-glass posts are sometimes used for this purpose for the same reason. The most recent use for ceramics in dentistry is orthodontic brackets. The development and demand for these items has been driven solely by aesthetics. Polycrystalline alumina is the material of choice in this application.

Electrical Engineering and Electronics

Products made of technical ceramics are now proven components in the construction and control of sophisticated plants, machinery and equipment with electrotechnical component assemblies.

Often, they make possible the function of such constructions in the first place. Typical examples include λ sensors in automotive engineering or in kiln and furnace engineering, the vacuum chambers of particle accelerators or actuators in motion detectors. The size of such components is typically in the region of a few millimetres up to several metres.

A special feature of this class of materials is the wide range of electrical conductivity, which spans more than 15 orders of magnitude and cannot be matched by any other class of materials. It includes electrically insulating as well as semi-conducting, ionic-conducting and superconducting materials. On top of this come the dielectric properties, which can be used, for instance, in sensor technology and telecommunications.

Besides the electrical properties, magnetic properties are often required. The soft or hard magnetic ferrites have proven effective materials for decades. Compared with metallic materials, they often enable smaller product sizes and therefore more economically attractive products.

Typical for the applications of technical ceramic materials is a frequent requirement for other non-electrical properties such as:

- Mechanical strength;
- Thermal resistance;
- Thermal shock resistance;
- Thermal conductivity;
- Corrosion resistance;
- Production of ultrahigh-vacuum-capable joints with metals.

A key strength of this class of materials is the demand-driven combinations of the above-mentioned properties. In addition comes the possibility to optimize properties for a specific application by means of appropriate doping and therefore to tailor materials to requirements.

For instance, the zirconia ceramics used for λ sensors can be optimized to maximize their strength without compromising their suitability for use as oxygen sensors as only one electrical signal must be recorded, which can be evaluated on the basis of appropriate calibration. With the selective modification of the chemical composition of this material, maximized electrical conductivity can be obtained along with an acceptable level of strength. As a result, this anionic conductor can be efficiently used as electrolyte for SOFCs (high-temperature fuel cells).

Doping of the electrically insulating Al_2O_3 with sodium forms a superionic conductor for Na^+, which can be used as a separator in Na/S high-energy batteries.

Al_2O_3 ceramic with purity up to > 99.9% remains one of the most commonly used oxide ceramic electrical insulating materials even today. Its specific electrical resistance reaches 10^{16} Ω cm at room temperature, while at 1600 °C it still reaches a level of 10^6 Ω cm. For this reason, these materials have been proven components for electrical insulation especially in high temperatures for decades.

The electrical properties of technical ceramics in combination with their mechanical properties generally enable a reduction in the sizes of existing systems, often simplified designs and therefore significantly enhanced performance. Typical examples are electrical feedthroughs on Al_2O_3 basis to withstand pressure above 1000 bar or substrates made of AlN ceramic which thanks to their high thermal conductivity of more than 200 W $(m K)^{-1}$ are used in demanding electronic applications.

The wide-ranging application of technical ceramic materials in electrical engineering and electronics has led over the past decades to a correspondingly high number of variants in electrically passive and active materials. Today a broad spectrum exists including insulators, dielectrics, piezoelectrics, NTC and PTC ceramics, varistors, cationic, anionic and electron conductors on oxide and non-oxide basis, superconductors as well as soft and hard magnetic ferrites.

The use of ceramic components in electrotechnical assemblies often demands material-to-material joints with one or more metallic components that can comprise different materials. Such joints are usually associated with the demand for high or ultrahigh vacuum tightness and acceptable mechanical strength for use in the field. Such ceramic metal joints are generally realized by hard brazing based on MoMn procedures. For this purpose a thick film of molybdenum applied and then fired onto the ceramic is joined with the respective metal part with a silver-based solder as standard or, if corrosive attack is expected, with gold-based solder. This process can be generally applied for all types of Al_2O_3 ceramics, but only certain types of ZrO_2 ceramics. Several decades ago, an active soldering process was developed that enables direct soldering of metal components with ceramics without prior metallization. This process is applied today as an alternative to metallized ceramics, mainly for ZrO_2. With both processes, given appropriate design and selection of the ceramic and metallic materials, joint strengths exceeding 200 MPa are achieved.

References

- Ceramic-engineering, entry: newworldencyclopedia.org, Retrieved 14 July, 2019
- Ceramic-Technology-(SS), 8, Engineering-branches: myklassroom.com, Retrieved 11 January, 2019
- Ceramics, technology: britannica.com, Retrieved 18 April, 2019
- Ceramic-matrix-composites, basics-design: machinedesign.com, Retrieved 27 May, 2019
- Glossary-3d-printing-ceramics, glossary, 4sight: digitalfire.com, Retrieved 21 July, 2019
- Nanoceramics-nanotechnology, nanoscience-and-nanotechnology: what-when-how.com, Retrieved 13 July, 2019
- Ceramic-capacitor-working-construction-applications: elprocus.com, Retrieved 24 March, 2019
- Advanced-ceramics-in-the-aerospace-industry: ceramicindustry.com, Retrieved 1 June, 2019

Ceramics Forming Processes

There are numerous processes which are used in the formation of ceramics such as firing, extrusion, slip casting, dry pressing, ceramic mold casting, ceramic injection molding, tape casting and freeze casting. The topics elaborated in this chapter will help in gaining a better perspective about these ceramics forming processes.

Firing Process for making Ceramics

Firing clay transforms it from its humble, soft beginnings into a new, durable substance: ceramic. Ceramics are tough and strong and similar in some ways to stone. Pieces of pottery have survived for thousands of years, all because clay met fire.

The temperature needed to transform soft clay into hard ceramic is extremely high and is usually provided by a kiln. You cannot fire pottery in a home oven because ovens do not get up to the high temperatures of more than 1,500 °F that you need for firing clay.

Firing is the process of bringing clay and glazes up to a high temperature. The final aim is to heat the object to the point that the clay and glazes are "mature"—that is, that they have reached their optimal level of melting. To the human eye, pots and other clay objects do not look melted; the melting that occurs is on the molecular level. This process is usually accomplished in two steps: bisque firing and glaze firing.

Bisque Firing

Bisque firing refers to the first time newly shaped clay pots, or greenware, go through high-temperature heating. It is done to vitrify, which means, "to turn it glasslike," to a point that the pottery can have a glaze adhere to the surface.

Greenware is fragile. To start, it must be bone-dry. Then, it must be loaded into the kiln with a great deal of care. The kiln is closed and heating slowly begins.

A slow temperature rise is critical. During the beginning of the bisque firing, the last of the atmospheric water is driven out of the clay. If it is heated too quickly, the water turns into steam while inside the clay body, which can cause the clay to burst.

Clay Transformation in the Bisque Firing

When a kiln reaches about 660 °F, the chemically bonded water will begin to be driven off. By the time the clay reaches 930 °F, the clay becomes completely dehydrated. At this point, the clay is changed forever; it is now a ceramic material.

The bisque firing continues until the kiln reaches about 1730 °F. At this temperature, the pot has sintered, which means it has been transformed to the point that it is less fragile while remaining porous enough to accept the application of glazes.

After the desired temperature has been reached, the kiln is turned off. The cooling is slow to avoid breaking the pots due to stress from the temperature change. After the kiln is completely cool, it is opened and the newly created "bisqueware" is removed.

Glaze Firing

Ceramic glaze is an impervious layer or coating applied to bisqueware to color, decorate, or waterproof an item. In order for earthenware, like fired clay pottery, to hold liquid, it needs a glaze. Potters apply a layer of glaze to the bisqueware, leave it to dry, then load it in the kiln for its final step, glaze firing.

The glazed item is carefully loaded into the kiln for the glaze firing. It must not touch other pots, or the glazes will melt together, fusing the pots permanently. The kiln is heated slowly to the proper temperature to bring the clay and glazes to maturity, then it is slowly cooled again. The kiln is opened and unloaded after it has cooled completely.

This second kiln firing causes a remarkable change in the clay and glaze. It completes the transformation of pots from a soft, fragile substance to one that is rock-hard and impervious to water and time.

Sintering

Sintering (Firing) of ceramic materials is the method involving consolidation of ceramic powder particles by heating the "green" compact part to a high temperature below the melting point, when the material of the separate particles difuse to the neighboring powder particles.

The driving force of sintering process is reduction of surface energy of the particles caused by decreasing their vapor-solid interfaces. During the diffusion process the pores, taking place in the "green compact", diminish or even close up, resulting in densification of the part, improvement of its mechanical properties.

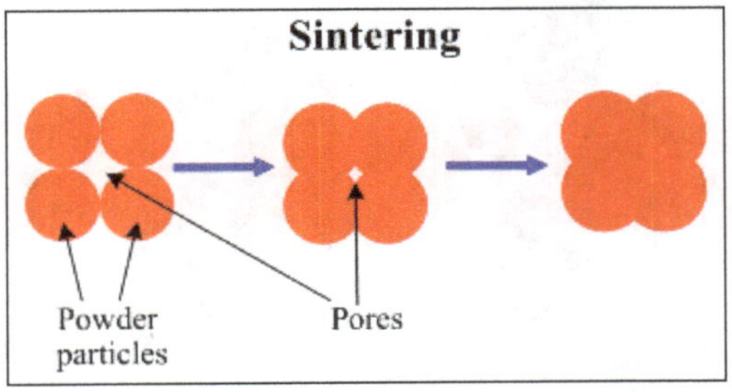

Decrease of the porosity, caused by the sintering process, is determined by the level of the initial porosity of the "green" compact, sintering temperature and time. Sintering is enhanced if a liquid phase takes part in the process (liquid phase sintering).

Sintering (firing) of pure oxide ceramics require relatively long time and high temperature because the diffusion proceeds in solid state. Applying pressure decreases sintering time and the resulted porosity.

Tunnel kilns and periodic kilns are commonly used for ceramics sintering (firing). In periodic kilns heating and cooling sintering stages are conducted according to a prescribed procedure. In tunnel kilns the sintered parts are conveyed through different temperature zones.

Typical tunnel kiln has three zones:

- Preheat zone for removing lubricant and other organic materials;
- Sintering zone where the diffusion occurs;
- Cooling zone where the sintered parts cool down.

Sintering process may be conducted in different atmospheres: air, inert atmosphere. Sintering occurs simultaneously with pressing in the hot pressing processes (hot die pressing, hot isostatic pressing).

Spark Plasma Sintering

Spark plasma sintering (SPS) or pulsed electric current sintering (PECS) is a sintering technique utilizing uniaxial force and a pulsed (on-off) direct electrical current (DC) under low atmospheric pressure to perform high speed consolidation of the powder.

Spark Plasma Sintering Mechanism

The mechanism of the spark plasma sintering is still unclear therefore the process name pulsed electric current sintering (PECS) should be considered as more correct. However commercially the name SPS (spark plasma sintering) is more common. The most accepted mechanism of the pulsed electric current sintering (PECS) is based on the micro-spark discharge in the gap between neighboring powder particles.

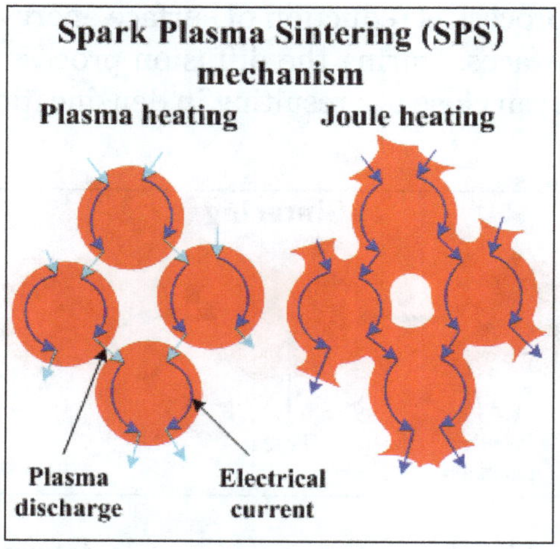

The spark plasma sintering process proceeds through three stages:

1. Plasma heating: The electrical discharge between powder particles results in localized and momentary heating of the particles surfaces up to several thousand °C. Since the micro-plasma discharges form uniformly throughout the sample volume the generated heat is also uniformly

distributed. The particles surfaces are purified and activated due to the high temperature causing vaporization of the impurities concentrated on the particle surface. The purified surface layers of the particles melt and fuse to each other forming "necks" between the particles.

2. Joule heating: At this stage the pulsed DC electrical current flows from particle to particle through the necks connecting them. The joule heat is generated by the electrical current. The joule heat increases the diffusion of the atoms/molecules in the necks enhancing their growth. Localized character of heating and its uniform distribution allow rapid temperature rise and drop, which diminish coarsening (growth) of the material grains.

3. Plastic deformation: The heated material becomes softer and it exerts plastic deformation under the uniaxial force. Plastic deformation combined with diffusion result in the densification of the powder compact to over 99% of its theoretical density.

Spark Plasma Sintering Technique

Spark plasma sintering (SPS) is performed in a Graphite die. The mechanical scheme of the process is similar to the Uniaxial (Die) Pressing. The load (commonly up to 15000 psi/100 MPa) is transferred to the powder through the upper punch. The pulsed DC power supply is connected to the upper and lower punches/electrodes.

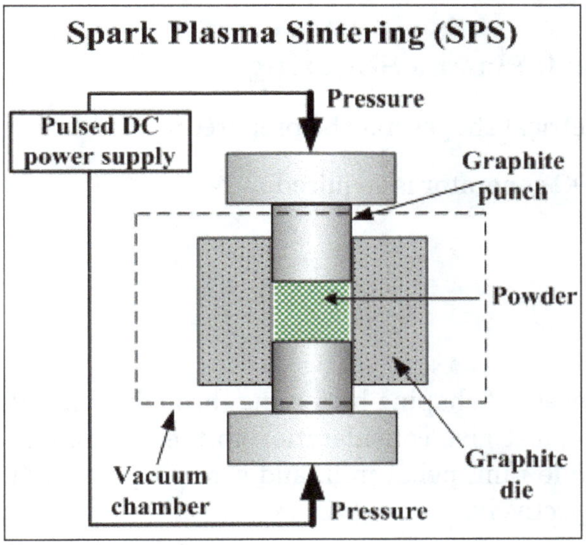

The pulsed DC electrical current parameters:

- Voltage: Few volts;
- Current: Several thousand amperes;
- DC pulse time: Few to tens of ms;
- DC pause time: Few to tens of ms.

The process is conducted under either low pressure (vacuum) or inert gas atmosphere. Spark plasma sintering process is fast. Its overall duration is commonly 5-20 minutes.

Since only surface layers of the powder particles are heated in the spark plasma sintering process, the average (monitored) temperature of the compact is relatively low: few hundred °C lower than in conventional sintering process.

Advantages of Spark Plasma Sintering

- Fast sintering process;
- Uniform sintering;
- Low grain growth (nano-grain materials may be prepared);
- Compaction and sintering stages are combined in one operation;
- Binders are not necessary;
- Better purification and activation of the powder particles surfaces;
- Different materials (metals, ceramics, composites) may be processed;
- High energy efficiency;
- Easy operation.

Disadvantages of Spark Plasma Sintering

- Only simple symmetrical shapes may be prepared;
- Expensive pulsed DC generator is required.

Extrusion

Extrusion is "the act or process of shaping by forcing through a die." When pelletizing or specialty-forming a product, one must give consideration to the advantages of extrusion. The various methods of extrusion include ram, pellet mill, and screw type, and "force" is applied by piston, rolls, or screw (auger), respectively.

A multiple-hole plate is commonly used in pelletizing.

A ram extruder is a piston inside a cylinder. The piston must be retracted to charge the hopper before it can be extended into the barrel. Once charged, material is moved by the piston or plunger into the barrel and compressed against the die, creating pressure. Ram extrusion is intermittent due to this basic functionality.

A pellet mill typically employs two rolls inside a cylinder. The rolls are attached to an arm that rotates on the axis of the cylinder, and this causes the rolls to rotate against the cylinder's inner diameter where the die plates are mounted. Product is introduced to the cylinder and forced through the dies with this rolling action. Pellet mills deliver continuously, but equipment size becomes an issue when significant throughput is required.

Screw-type extrusion incorporates an auger inside a feed hopper, which continues into a barrel. When material is introduced to the hopper, it is conveyed by the screw into the barrel and compressed against the die, creating pressure. It differs from the ram in that the screw rotates continuously to create a constant flow of material. The continuous nature of screw-type extrusion offers maximum productivity through the inherent design. However, many factors must be considered in order to maximize the extrusion process.

Sizing and Options

When selecting equipment, it is important to know the throughput requirements, space constraints, input power requirement and profile specifications. The equipment manufacturer will also require product information and may request a sample of material for laboratory testing. This information is crucial in order to determine the appropriate equipment size, feed requirements, and screw and die design.

Once the extruder is sized, drive options should be taken into consideration. An overall compact design will afford more flexibility, whether designing a new facility or fitting a new line into an old facility, and various drive options add to this flexibility. A gear motor is the most compact and requires less maintenance than belts and sheaves, but space constraints sometimes dictate design. Where D/C used to be the standard, A/C motors and inverter controllers are now more common as constant torque is available at a fairly low cost in ratios ranging from 3:1 to 10:1. Though rarely required, ratios up to 1000:1 are available at a slightly higher premium.

A forming die can range from a very simple, solid cylindrical shape to an extremely complex monolithic shape.

Ancillary equipment as it relates to the desired end product also needs to be considered. In some pelletizing applications, it is sufficient to simply extrude onto a conveyor belt. The strands will eventually break into lengths that are acceptable. If size requirements are more stringent, it is necessary to cut the extrudate. Various means of cutting are available in forming applications, including die face, guillotine, traveling cutters, etc., and all are dependent on the specific application.

A multiple hole plate is commonly used in pelletizing, while a forming die can range from a very simple, solid cylindrical shape, such as that used in electrical insulator pugs, to an extremely complex monolithic shape like those of automotive catalytic converters. More complex shapes are not only more costly to manufacture but also require improved flow properties that sometimes necessitate the use of extrusion aids.

Die design is as important to the quality of the extrudate as it is to throughput of the extruder. Open area, land area, and material of construction directly relate to back pressure, and in turn, product density and screw efficiency. Open area refers to the actual cross-section of the die and is typically expressed in ratio to the cross section of the screw. The land area of the die is the portion that forms the shape of the product. Material of construction will affect pressure through friction. Carbon steel, for example, has a much higher coefficient than that of acetyl plastic.

Flow characteristics of the extruded material also impact screw and die design. A product with the consistency of bread dough will flow quite easily through any orifice. However, a material that is more like corn meal will be more difficult to process. Extrusion aides ranging from petroleum to soy to clay are available to increase a product's lubricity and ease the extrusion process.

Design Considerations

The overall design of the extruder with respect to how the material passes through is vital. For example, the buildup of residual material in the feed hopper is a common problem in the extrusion pelletizing of certain catalyst carriers. The buildup is harmless until it dries out, breaks loose and causes die plugging, which is costly because it reduces throughput and eventually requires downtime for die removal/replacement.

One solution to die plugging is to utilize an improved feed hopper design that includes counter-rotating feeders to keep the material from bridging and from building up on the hopper walls. These feeders are configured above the screw in a manner that directs material into the screw efficiently, and they are designed to be virtually self-cleaning.

Another important factor is the clearance between screw and barrel. As clearance increases, the actual volume delivered per revolution decreases, along with die pressure. Conversely, a drop in die pressure can be an indication of increased clearances. It is important to monitor wear to ensure maximum efficiency. Tighter clearances mean higher efficiency, and maintaining a tight tolerance is crucial to maintaining output. Tighter tolerances, along with other design improvements, can be implemented to prolong the life of the barrel and screw. The advantages are maintained productivity, less downtime, reduced rebuild costs and less die plugging.

Temperature is important to extrusion for a number of reasons relative to each product or process. When designing cooling channels, it is therefore imperative to utilize as much surface area as possible and create flow patterns that transfer heat more efficiently. This is achieved through improved barrel, barrel jacket and screw design. Giving more attention to DT, dead spots and turbulence can greatly increase cooling efficiency.

Self-monitoring Systems

In the past, much of what determined the performance of an extrusion system were the individuals who controlled the process. The operators would determine the consistency of the mix, feed levels, extruder speeds based on their particular knowledge base and level of experience. Although some of these individuals deserve much commendation for the results they consistently obtained, others might not achieve the same quality or performance, thus creating a perpetual QC headache.

It is often better to limit human involvement and utilize a system that is self-monitoring and features the ability to adjust itself accordingly. Today's electronics make these systems both practical and affordable. For instance, a mechanical feeder can be supplied that operates on an electronic feedback loop. Once the desired output is achieved, the extruder speed is set and the feeder adjusts itself to maintain a desired hopper level. The advantage is a more steady extrusion rate, more stable die pressure and consistent product density.

Counter-rotating feeders can keep material from bridging and from building up on the hopper walls.

Slip Casting

Slip casting facilitates the mass producing of ceramics, but doesn't need any particularly specialist equipment and can be done in the studio. Dependent on how many molds you have, slip casting can create hundreds of identical pieces. Some potters tend favor this method, as slip casting makes pieces that are tricky to hand build or create on the wheel much easier. Another brilliant benefit of slip casting is that you can make wares like a jug with a handle as a single piece. Slip casting is also a great method to create more complicated pieces, such as a teapot.

Slip is just a liquid mixture of clay or a "liquefied suspension of clay particles in water" and is generally a little thinner than slurry. Slip casting is "liquefied clay poured into a plaster mold." The ratios are roughly 75 percent clay body to 25 percent water.

Turn the Clay Body into Slip

The process of transforming the clay into a slip is called the slaking process. It's important to use the same types of clay so it can be successfully fired at the same temperature. Work out the quantities of clay and water you need and add them to a large bucket, leaving them to slake (disintegrate). Different clays have different slaking times, but you should leave the clay to dissolve in the water for around 24 hours. The clay should be bone dry for this process. After this time, the mixture will need to be stirred gently with a large stick. To create the ideal consistency (the slip should be thickness of cream), you'll need a drill with a mixing attachment to make sure the slip is smooth. It's important to work out if the slip has the correct clay to water ratio (known as its specific gravity). To measure the specific gravity, you'll need to pour 100 ml of the clay slip into a measuring jug and weigh it in grams. Then divide the weight you get by 100 (or the amount of milliliters you have poured out) and that number will be the specific gravity. A good ratio to work with is roughly 1.80. Then you'll need to make sure the clay is deflocculated.

Deflocculation

Chemicals need to be added to the clay slip so that it has the correct fluidity. This is known as deflocculation. Deflocculation is to "disperse the particles in a slip so that less water is required to make the slip fluid." A few drops of sodium silicate can be used to achieve this. A liquid deflocculant like Darvan is also good, as it will not erode a plaster mold. Wear a respirator mask throughout the whole process of turning the clay body into slip, so as you do not inhale any harmful chemicals.

Pour out the Casting Slip

Once the slip is the correct consistency, then you'll need to have the right plaster mold ready. Plaster molds are readily available to buy or you can follow our tips and tricks to making the own plaster mold. Molds are made from plaster of Paris as they're highly absorbent.

Carefully pour the slip into the cavity of the mold and the plaster of Paris mold will start to absorb water from the clay and the drying process of the clay will begin. Once the clay has started to dry to a suitable thickness (the timing will depend on the size of the mold, but for a medium/small mold, 10 minutes should do), then you can pour out the excess clay slip. There is a knack to pouring out the clay slip and you'll need to do it very carefully. Pour it back into the bucket of the clay slip, so as not to waste any. When pouring, put two pieces of wood above the bucket to rest the plaster of Paris mold on, upside down, so the slip pours out evenly. Some potters recommend pouring the slip out at an angle, so that the inside is even smoother. Once it's all poured out, and then slice off any excess clay from the top of the mold.

Tipping the Ware out of the Mold

Once the piece is dry, (the length of time the clay takes to dry will be dependent on the size of the mold you use) you can tip it out. Place a flat board underneath the mold and carefully turn it over. This is the easiest method and the piece should gently come out, as the clay will have shrunk in the mold. From there you can tidy up the edges and the piece will be ready for firing and glazing.

Dry Pressing

Dry pressing ceramic is the most economic process for large production runs, and is suitable for both simple and complex geometries. Depressions and holes are normally only designed in the pressing direction.

Dry pressing is characterized by very slight residual moisture. It is almost non-existent. Any further drying is not needed. The process is thus eminently suitable for highly dimensioned accurate parts in large quantities. Practically all ceramic materials can be deployed.

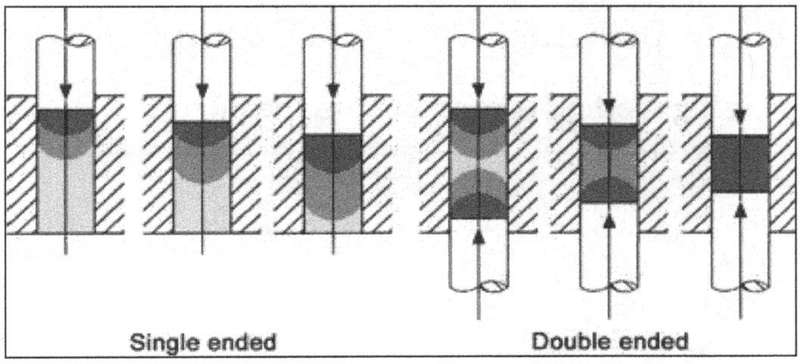

A pressing tool for axial dry pressing consists of a press die as well as top and bottom rams. Furthermore, various subdivisions of the die are possible depending on component complexity. A filling shoe is used for filling the press granulates into the mold. With two-sided pressing, a controlled movement of top and bottom rams results in compaction of the granulate. With one-sided pressing only the upper die is in motion. The compact is automatically extracted via pusher, gate and gripper. Dry Pressing is used to manufacture mass-produced precision ceramic products. Non-clumping granulates are compressed in steel dies designed appropriately for the part to be manufactured. The high cost for the dies can only usually be justified for large runs.

Depending on the design of the ceramic dry pressing machine, components ranging in size from tiles down to match heads can be manufactured. Small discs or plates can be pressed with thicknesses of around 0.8 or 1.0 mm. The tape casting process is more suitable for even thinner, flatter components. It is still possible to manufacture fine ridges or similar structures on the component if the granulate being pressed can effectively fill hollows in the pressing tool, and provided it is possible to create the necessary tool.

Methods of Shape Forming Ceramic Powders

Prior to shape forming ceramic powders are mixed with processing additives (binders, plasticizers, lubricants, deflocculants, water). Powder preparation stage is followed by shape forming stage.

The following techniques are involved in forming ceramic powders into a desired shape:

- Uniaxial (Die) Pressing,
- Isostatic Pressing,
- Injection Molding,
- Extrusion,
- Slip Casting,
- Gel Casting,
- Tape Casting.

Uniaxial Pressing

Die pressing is the powder compaction method involving uniaxial pressure applied to the powder placed in a die between two rigid punches. Uniaxial (die) pressing is effectively used for mass production of simple parts (alternative method is isostatical pressing).

The method of the die pressing method is presented in the picture.

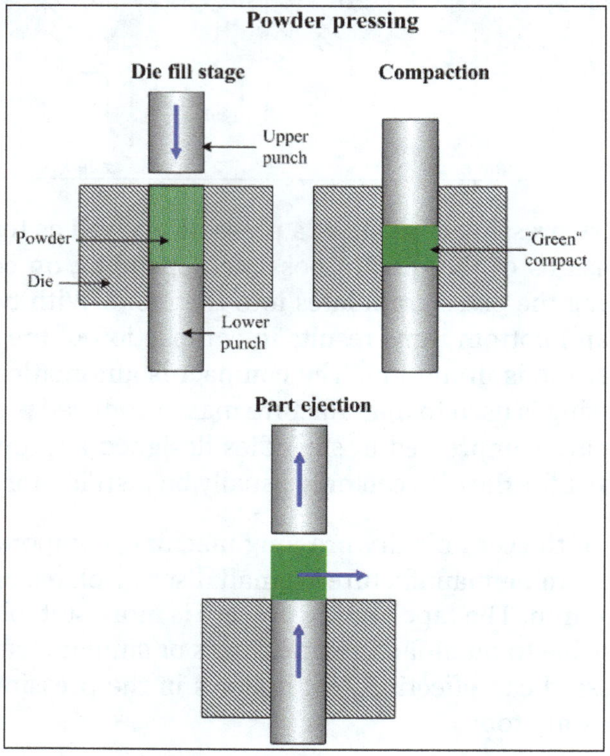

The pressing process consists of the following stages:

- Die filling: At this stage a controlled amount of the powder is fed into the die cavity.
- Compaction: Upper punch moves down and presses the powder with a predetermined pressure. The pressure varies between 10,000 psi to 120,000 psi (69 MPa to 830 MPa).
- "Green" compact part ejection and removal ("green" compact – unsintered powder compact).

The pressing cycle repeats 400 to 5000 times/hour, depending on the press type, powder filling properties and the part size and geometry.

Hydraulic and mechanical presses with load up to 750 tons are used for the powder die pressing. Die pressing, which is conducted at thr room temperature is called cold pressing. If the pressing process is conducted at increased temperature it is called hot pressing.

For example hot pressing of silicon carbide is carried out at temperature about 3630 °F (2000 °C). Hot pressing permits obtaining better compaction, higher green compact density and higher (than in cold pressing) strength of the part. Die pressing is used for manufacturing insulating parts, magnetic ceramics, capacitors, substrates.

Isostatic Pressing

Isostatic pressing is the powder compaction method involving applying pressure from multiple directions through a liquid or gaseous medium surrounding the compacted part. Cold isostatic pressing (CIP) is conducted at room temperature.

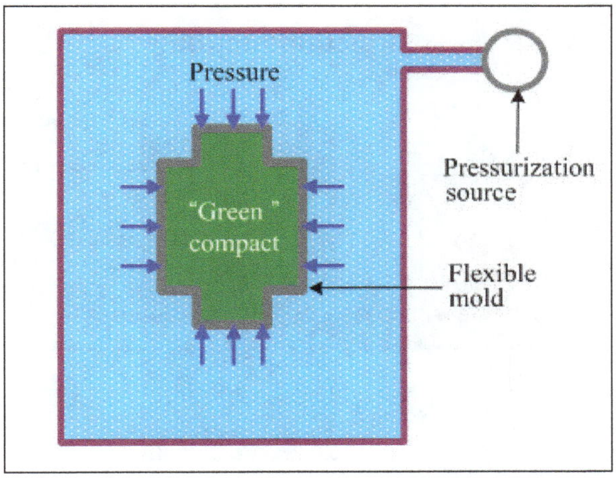

Cold isostatic pressing.

A flexible (commonly polyurethane) mold immersed in a pressurized liquid medium (commonly water) is used in the cold isostatic pressing method. There are two types of cold isostatic pressing: wet bag and dry bag.

- In the wet bag method the mold is removed and refilled after each pressure cycle. This method is suitable for compaction of large and complicated parts.
- In the dry bag method the mold is an integral part of the vessel. The dry bag method is used for compaction of simpler and smaller parts.

The cold isostatic pressing (CIP) method has the following advantages as compared to the die cold pressing method:

- Better uniformity of compaction.
- More complex forms (for example long thin-walled tubes) may be compacted.

Hot isostatic pressing (HIP) involves isostatic pressing conducted at increased temperature. As a pressure medium a gas (Nitrogen or Argon) is used. The work pressures, which are applied in the hot isostatic pressing method, are commonly between 15,000 psi to 44,000 psi (100 MPa to 300 MPa).

Hot isostatic method (HIP) combines pressing and sintering, causing consolidation of powder particles, healing voids and pores. The part shrinks and densifies, forming sound high strength structure. Cans made of stainless steel, or mild steel are used in the method.

The method may be used without a mold. In this case the part is first compacted by cold isostatic pressing method, and then it is sintered in order to close the interconnecting porosity. The sintered (but still porous) part is then pressed isostatically at high temperature without any can (mold).

Injection Molding

Injection moldingis the method of compaction of ceramic powder fed and injected into a mold cavity by means of a screw rotating in cylinder. The method is similar to the plastic injection molding. The principal scheme of the injection molding method is shown in the picture.

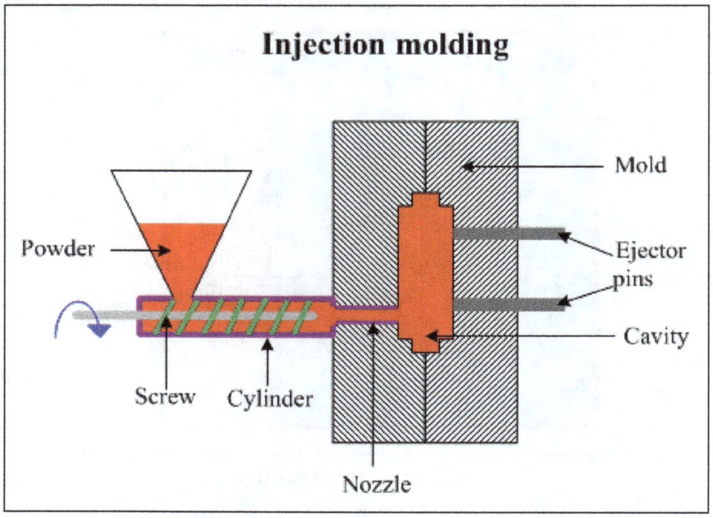

The processing technology comprises the following stages:

- Mixing the ceramic powder with 30% - 40% of a binder – low melt polymer.
- Injection of the warm powder with molten binder into the mold by means of the screw.
- Removal of the part from the mold after cooling down of the mixture.
- Debinding – removal of the binder. There are two debinding methods:
  - Solvent debinding – the binder is dissolved by a solvent or by water;
  - Thermal debinding – the binder is heated above the volatilization temperature.
- Sintering the “green” compact.

Injection molding method is widely used for manufacturing small parts having complex shapes. The molding cycle in the injection molding method is about 10 sec, which much less, than the molding time in the alternative methods – 10-20 min. The method permits to produce part with close tolerance, due to the consistent shrinkage. This shrinkage is taken into account in the mold design.

Extrusion

The two possible schemes of extrusion are presented in the picture. Extrusion ram forces the ceramic paste through a die, resulting in a long product (rods, bars, long plates, pipes) of regular cross-section, which may be cut into pieces of required length. Extrusion is used for manufacturing furnace tubes, thermocouple components, and heat exchanger tubes.

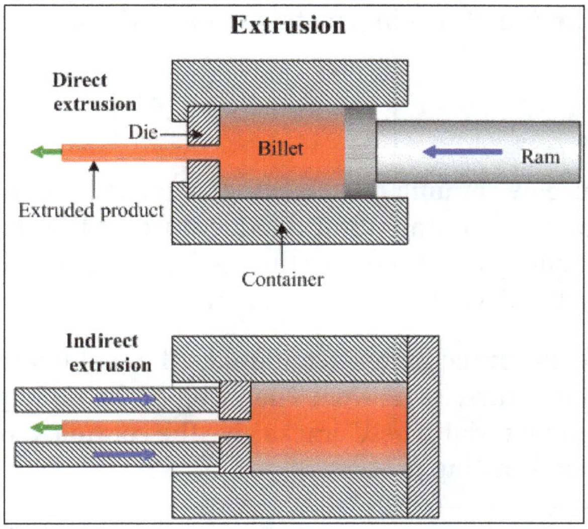

Slip Casting

Slip casting technique is used for shape forming ceramic slurry - a stable suspension, consisting of ceramic powders, processing additives and 20-35% of liquid (water or solvent). The slurry is poured into a mold made of a micro-porous material (e.g. gypsum), which is capable to soak the liquid from the slurry, consolidating the ceramic part. Slip casting is used for manufacturing fine china, sinks, sanitary ware, thermal insulation parts.

Gel Casting

Gel Casting is a process of shape forming slurry prepared from ceramic powder mixed with a solution of organic monomer. When the slurry is poured into a mold, the monomer polymerizes, forming a gel that binds ceramic powder particles into a strong complex-shaped part. The parts may be machined before firing. The process is economical and it is used for manufacturing large complex shapes parts such as turbine rotors.

Tape Casting

Tape Casting is a process of forming a thin film of ceramic slurry spread over a flat surface. The slurry film thickness is controlled by a knife edge located above the moving carrier surface. The solvent, containing in the slurry, evaporates, resulting in formation of ceramic sheet, which may be stripped from the supporting surface. Tape Casting is used for manufacturing multilayer ceramics for capacitors and dielectric insulators.

Ceramic Mold Casting

The manufacturing process of ceramic mold casting is like the process of plaster mold casting but can cast materials at much higher temperatures. Instead of using plaster to create the mold for the metal casting, ceramic casting uses refractory ceramics for a mold material. In industry, parts such as machining cutters, dies for metalworking, metal molds, and impellers may be manufactured by this process.

Process

The first step in manufacture by ceramic mold casting is to combine the material for the mold. A mixture of fine grain zircon ($ZrSiO_4$), aluminum oxide, fused silica, bonding agents, and water, creates ceramic slurry. This slurry is poured over the casting pattern and let set. The pattern is then removed and the mold is left to dry. The mold is then fired.

The firing will burn off any unwanted material and make the mold hardened and rigid. The mold may also need to be baked in a furnace as well. The firing of the mold produces a network of microscopic cracks in the mold material. These cracks give the ceramic mold both good permeability and collapsibility for the metal casting process.

Once prepared, the two halves of the mold are assembled for the pouring of the metal casting. The two halves, (cope and drag section), may be backed up with fireclay material for additional mold strength. Often in manufacturing industry, the ceramic mold will be preheated prior to pouring the molten metal. The metal casting is poured, and let solidify. In ceramic mold casting, like in other expendable mold processes, the ceramic mold is destroyed in the removal of the metal casting.

Properties and Considerations of Manufacturing by Ceramic Mold Casting

- Manufacturing by ceramic mold casting is similar to plaster mold casting in that it can produce parts with thin sections, excellent surface finish, and high dimensional accuracy. Manufacturing tolerances between .002 and .010 inches are possible with this process.
- To be able to cast parts with high dimensional accuracy eliminates the need for machining, and the scrap that would be produced by machining. Therefore precision metal casting processes like this are efficient to cast precious metals, or materials that would be difficult to machine.
- Unlike the mold material in the plaster metal casting process, the refractory mold material in ceramic casting can withstand extremely elevated temperatures. Due to this heat tolerance, the ceramic casting process can be used to manufacture ferrous and other high melting point metal casting materials. Stainless steels and tool steels can be cast with this process.

- Ceramic mold casting is relatively expensive.
- The long preparation time of the mold makes manufacturing production rates for this process slow.
- Unlike in plaster mold casting, the ceramic mold has excellent permeability due to the micro-crazing, (production of microscopic cracks), that occurs in the firing of the ceramic mold.

Ceramic Injection Molding

Ceramic Injection Molding (CIM) is a manufacturing process used to mass produce precision ceramic components with complex geometries. The 4 basic processes of mixing, injection, debinding and sintering are similar to that of Metal Injection Molding (MIM).

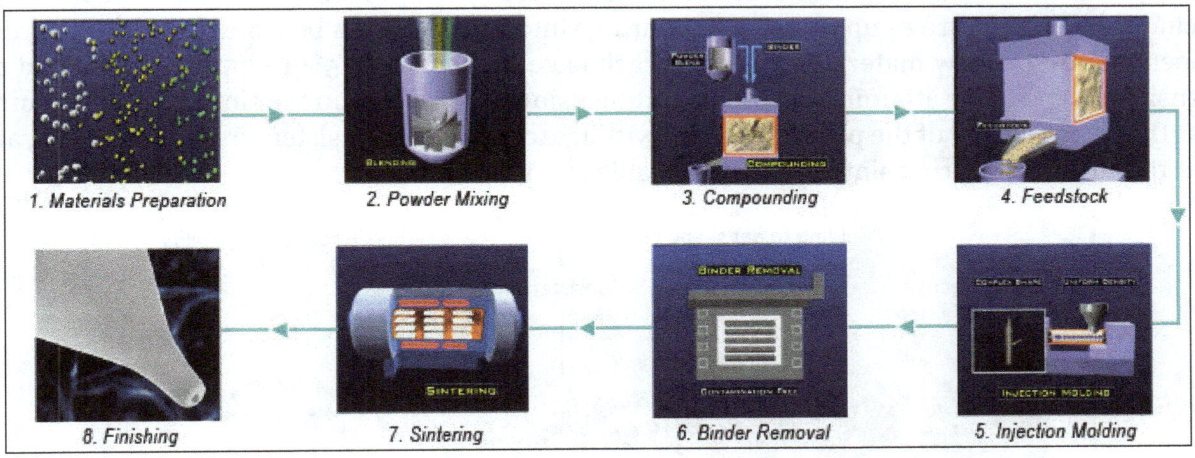

Process of Ceramic Injection Molding

Small Precision Tools' injection molding process is a combination of powder, injection molding, and sintering technologies. To obtain the necessary chemical and physical properties, powders are selected by size and shape and complemented with additives. Every particle of the powder is coated with binder components, which transport the powder for molding and gives the final form rigidity.

The ceramic injection molding is very suitable for high volume production of complex design with tight tolerances like bonding capillaries. It is an effective way of manufacturing complex precision components with the highest degree of repeatability and reproducibility.

Process Flow for Capillary Manufacturing Process

Stage 1: The characteristics of the ceramic powders, such as grain size distribution and morphology play a vital role not only in the achievement of the desired product properties, but also in the success of the different stages of the process. The ceramic powders used for the manufacturing of capillaries are selected by size and shape and complemented with additives to obtain the necessary chemical and physical properties.

Stage 2 – 3: Before the injection molding, the powder is mixed with binder to form a homogeneous mixture that is used to form the shape of the capillary. The binder is used for the artificial plasticization of the ceramic powders and for the formation of the desired shape through injection molding. Consideration for binder selection includes the flow characteristics for injection molding, the ease of binder removal and binder-powder interaction.

Stage 4 – 5: Subsequently, the mixture is feed to the mold for injection molding. The molding process is a significantly affected by the temperature, pressure and time envelope and it is essential to have the correct temperature and pressure sequence together with the time sequence.

Stage 6: The binder is removed by evaporation and exothermic reaction, leaving only a small fraction behind. Removal of the binder is a critical step between the molding operation and the sintering process. The extent of the binder removal requires careful monitoring and control to retain the shape of the capillary.

Stage 7: The formed part is then sintered in an oxidizing or reducing atmosphere, or in a high vacuum at temperature of up to 1800 °C. During sintering, the parts become denser and shrink. Depending on the raw material properties, shrinkage ranges from 15% to 25% of its molded dimensions. Repeatable attaining the required dimensional tolerances on the sintered parts requires that the green density of the part be uniform within each part and consistent from part to part, and that the shrinkage during sintering be repeatable and predictable.

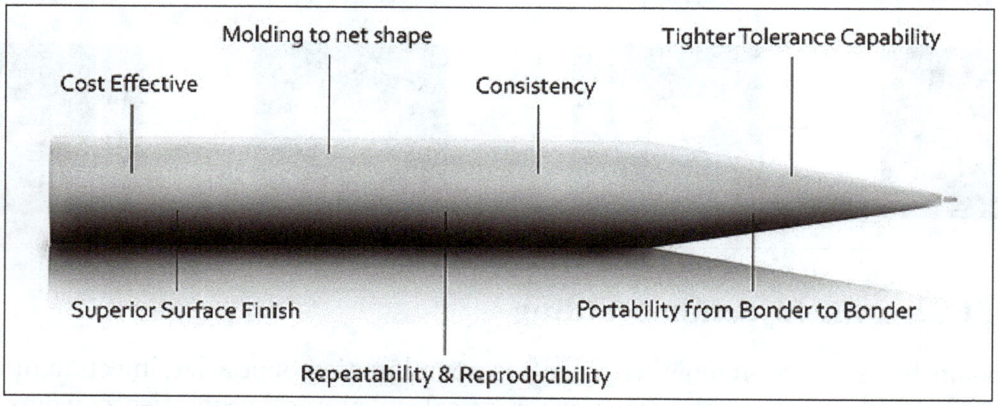

Advantages of Ceramic Injection Molding

Small Precision Tool's injection molding process offers a high degree of reproducibility. Complex parts in ceramic can be shaped in one operation with diverse geometry, different profiles, undercuts, sharp edges, and different wall thickness.

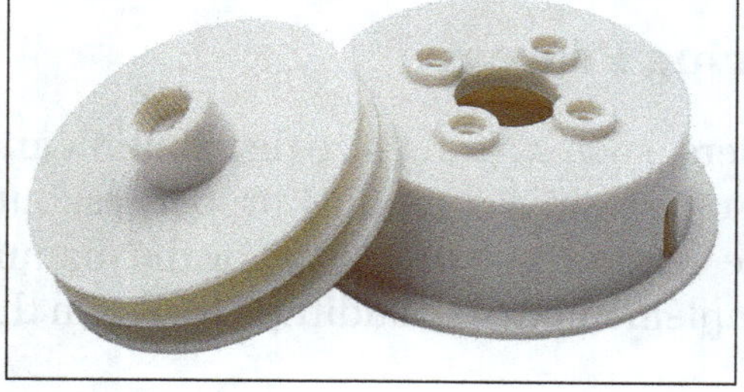

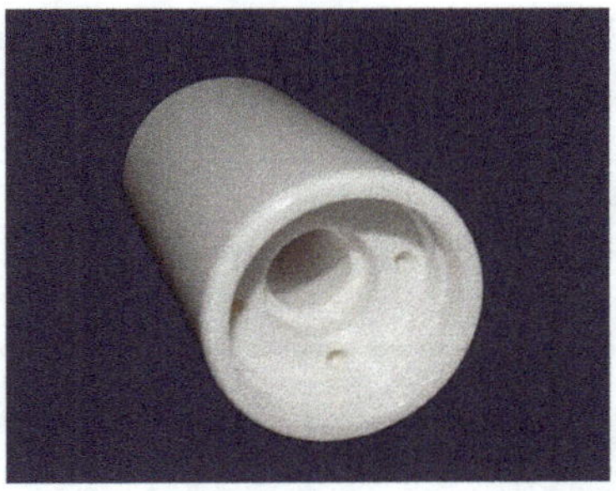

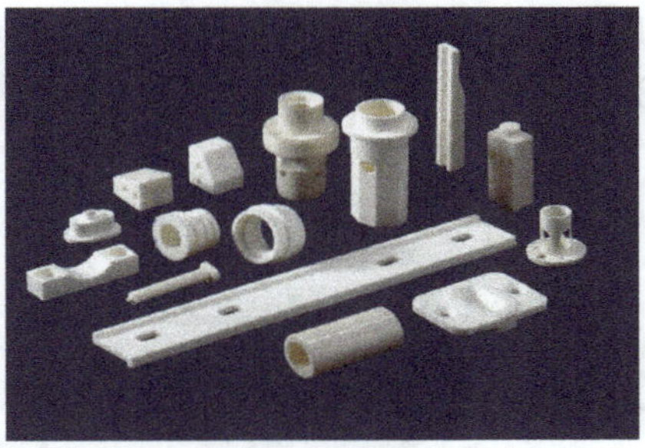

Precision Ceramic Parts.

Application Horizon

Today, the Small Precision Tool's injection molding process is applied in the instrumentation, textile, automobile, printing, electronic assembly, communications, aerospace, optical, medical, dental and chemical industries. Cost effective applications are found in relatively small parts demanding complex machining operations, and where volume production requires a large investment in machine tools.

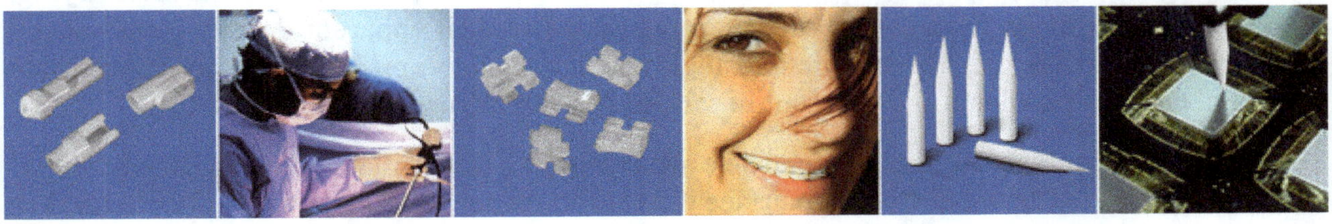

Tape Casting

Tape casting, often referred as doctor blading or knife coating, is widely used to form thin ceramic films mostly for capacitor and electronic packaging laminates. When combined with NP attributes, this technique could address current device miniaturization challenges including control of the green tape thickness. A scraping (or doctor) blade is used to remove excess deposited coating materials from a moving substrate. Film thicknesses from 5 μm to a few millimeters are currently obtained using tape casting. Additional post-processing steps are required to control film composition and properties. One clear advantage of tape casting is the possibility of using the continuous process on flexible substrates. Although tape casting is mostly based on non-aqueous solvent, aqueous-based tape-casting systems have also been used.

Higher ionic conductivity and improved mechanical properties are obtained by incorporating ceria gadolinia NPs as film in low-temperature solid oxide fuel cells (SOFC). Powder NPs (ceria gadolinia) were mixed with a solvent (butyl acetate), dispersant (sorbitane monooleate), plasticizer (poly-ethyleneglycol 300 and a bis(2-ethylhexyl) phthalate), and binder (poly-vinylbutanal) in well-defined proportions and mixing steps. The suspension was deposited onto a silicon-coated PET foil followed by a casting using a doctor blade with a slit width of 1 mm at a blade speed of 4 mm/s. After removal from the substrate foil, the casted film was heated up to 500 °C at a heating rate of 30 °C/h and held for 1 h to slowly remove all organic additives. It was then heated up at

30 °C/min to a final sintering step at 940 °C and 980 °C for 2 h. Following sintering, grain size increased from 37 to about 200 nm, with > 92% theoretical density.

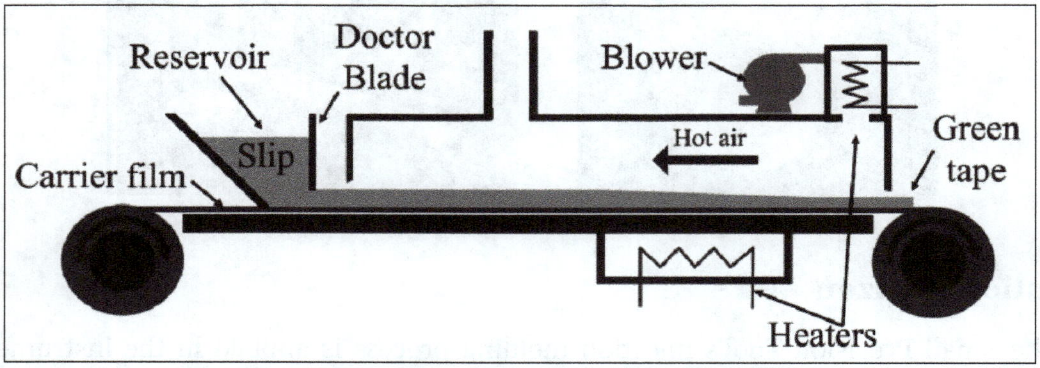

Simplified schematic outline of a continuous tape-casting setup.

For economical, toxicity, and environmental reasons, aqueous-based slurry for tape casting is becoming very attractive. The aqueous-based fabrication process is quite similar to non-aqueous tape casting. However, aqueous tape casting has some specific techniques such as slow drying, higher crack sensitivity, flocculation, and poor wetting. The high apparent density of the casted film requires relatively high inorganic and high plasticizer contents. Furthermore, appropriate milling of ceramic powder and dispersant in solvent is often required.

The relative NP content and the inter-particle separation have a dramatic effect on the quality of aqueous tape-casted films. Surface-to-surface separation distance (SSD) between NPs is used as the criterion to prepare crack-free NP-PVA (polyvinyl alcohol) films. Indeed, with the high Al_2O_3 NP content and small enough SSD, PVA binder is unable to bind the entire alumina NP inducing cracks within the casted film. In spite of all the effort, aqueous tape casting remains uneconomical. Although organic solvents are more expensive, low yield and slow processing time make aqueous tape casting still more expensive.

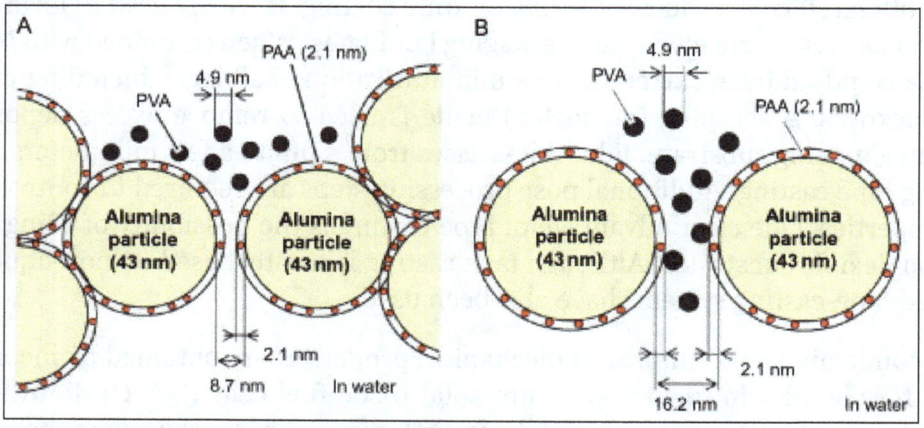

Schematic modelization of the effect of PAA-modified Al_2O_3 nanoparticle content on the interparticle separation.

Solvent type and dispersant are key components of tape-casting slurry. Dense and smooth green lead zirconate titanate $Pb(Zr_xTi_{1-x})O_3$ tapes were obtained using xylene–ethanol as solvent and phosphate ester as dispersant. The average starting particle size is about 75 nm. The doctor-blading technique, consisting of mechanically spreading a NP-based paste within a well-defined boundary, has been widely used to prepare 5–10 µm-thick TiO_2 porous films. Bodnarchuk et al. obtained square-inch assemblies consisting of ordered islands. Indeed, wustite/cobalt ferrite super-lattices are formed from magnetic 10–20 nm core/shell nanocrystals.

Tape casting is applied widely to producing functional ceramics (e.g., piezoelectrics, superconductors, and fuel-cell membranes), graded compositions, particulate and whisker-reinforced composites, and coatings. The technique is applied less frequently to continuous fiber composites and indeed subsequent to initial investigation and some later development in the mid-1990s little recent progress has been reported.

The underlying principle of tape casting is that a reservoir of slurry with a slit shaped outlet at the base is moved across a surface, releasing a layer of slurry. Slurry characteristics and the speed of movement of the reservoir can be adjusted to influence the thickness of the tape. A doctor blade can also be drawn along the tape surface to modify tape thickness. When the process is applied to fiber-reinforced ceramics, however, specific relationships between process parameters and tape thickness calculated for monolithic material cannot be applied directly. When applied to CMCs tapes, containing fibers and matrix powder are manufactured by tape casting matrix over an assembly of fibers. A suitable binder and plasticizer are included allowing the green preform to be assembled from stacks of tapes by pressing, sometimes with the application of heat. A burnout is carried out to remove binders, plasticizers, and surfactants and then densification is carried out at elevated temperature.

The distinctions between tape casting and the lay-up of tapes produced by other methods are somewhat blurred and the boundaries difficult to define. The initial patented work refers to Textron monofilament (140 µm diameter silicon carbide fiber deposited by CVD on a carbon core) reinforced reaction bonded silicon nitride. In fact, the composite tapes themselves are not tape cast. Rather, filaments are wound to a fixed spacing on a take up mandrel and held in place by coating with a fugitive polymeric binder. Monolithic tapes of silicon are produced by rolling dough-like slurry into sheets. Alternate layers of silicon and fiber mats are stacked, hot-pressed at 600 °C in air to remove binders, then vacuum hot-pressed at 1000 °C to consolidate the preform. This is then subjected to a separate nitridation step. This method is similar to the foil–fiber–foil methods used in the production of monofilaments titanium composites but due to a lack of plasticity in the matrix material during hot-pressing, some porosity remains between fibers, in the direction perpendicular to the hot-pressing axis.

Work by Holmquist et al. is more obviously tape casting. Lostec *et al.* used mechanically constrained Nicalon fibers on a flat surface and tape cast MAS-L ($MgO–Al_2O_3–SiO_2–Li_2O$) glass slurry as a matrix on top of them. A PVB (polyvinylbutyral) binder and DBP (dibuytylpthalate) plasticizer were used. Composites were densified by thermocompressing at 125 °C then hot-pressing at 1300 °C. Similarly, a mullite matrix was tape cast over arrays of Textron SiC monofilaments by. The monofilaments were wound on to a drum and then the arrays laid flat before slip casting. The method for securing the fibers is not explained. A latex binder was used in the high-purity mullite matrix slurry. Due to the low glass transition temperature of the particular latex binder used, lamination could be carried out at room temperature. These composites were densified by HIPing. It was shown here that processing at 1450 °C and 200 MPa or 1600 °C and 20 MPa pressure produces matrices of equivalent density. Fiber distribution was even and the matrix material was homogeneous with no pores or cracks between fibers. Mechanical properties of the composite processed at 1450 °C and 200 MPa were better than those of the material HIPed at 1600 °C. Finally, Pritchard and Shatwell dispensed with one of the lay-up steps by casting silicon slurry (PVB binder and PEG–polyethylene glycol–plasticizer) over a lay-up of Textron SCS-6 monofilaments on the tow laying mandrel. 0/0° and 0/90° laminates of the material were bonded at room temperature by rewetting the tape surfaces with solvent to redissolve the binder. Nitridation of this material was carried out at 1410 °C without the application of pressure. Careful control of fiber position during layup resulted in excellent fiber distribution. Matrix density was also high. Figure shows a SEM micrograph of a transverse section of a high-density unidirectional SiC monofilament-reinforced $MoSi_2$/Sialon matrix composite. This was produced via the same tape casting method. The composite was hot-pressed for an hour at 1750 °C under 25 MPa pressure.

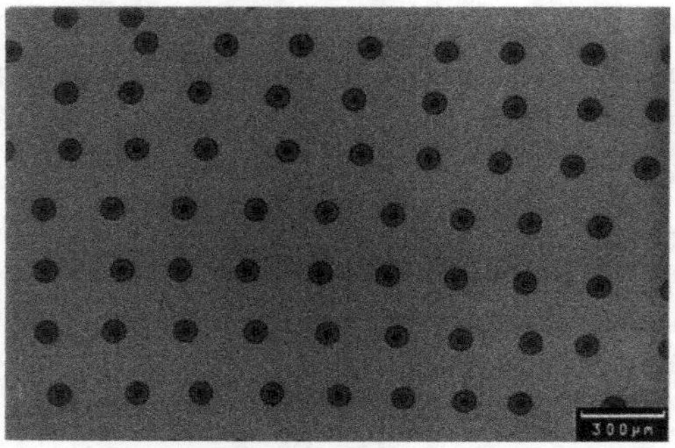

Hot-pressed SiC monofilament-reinforced $MoSi_2$/sialon material produced by tape casting and hot pressing.

Tape casting is most easily applicable to monofilament-reinforced ceramic composites. These do not compare favorably to materials reinforced with small diameter (~10 μm) fibers as the specific area of interface, and thus the scope for toughening is much reduced. The stacking method of production does however lend itself to the production of angle-ply laminates with simpler geometry than those materials produced from woven fibers.

Freeze Casting

Freeze casting is based on the freezing of a ceramic or hybrid aqueous suspension into a mould and the successive sublimation of the solidified solvent phase. Depending on the material typology a sintering process can be performed to consolidate the structure.

The porosity characterizing the material after the sublimation process is the replica of the ice crystals. By controlling the freezing, hierarchical architectures with anisotropic porosity can be realized which show porosity mainly formed by unidirectional channels separated by ceramic plates. This makes more effective the eventual successive infiltration processes with other phases.

Main factors influencing the process and the architecture of the porous product:

- The suspension concentration, usually in the range 10-40 vol%, inversely determines the total porosity percent.

- The freezing rate influences the globular/lamellar morphology and distribution of the pores.
- The particles dimensions of the solid in the suspension.

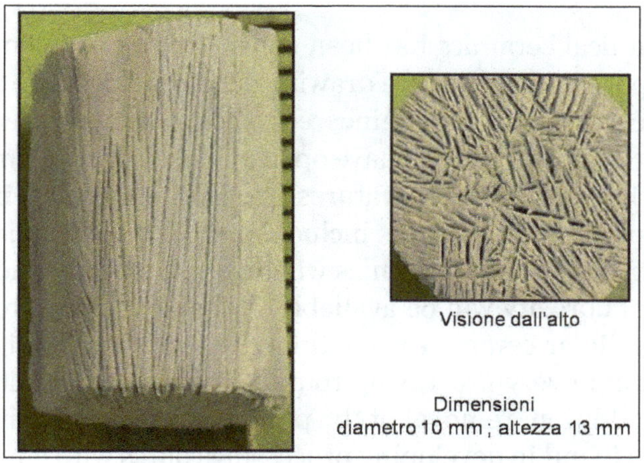

Shows a basic freeze casted samples.

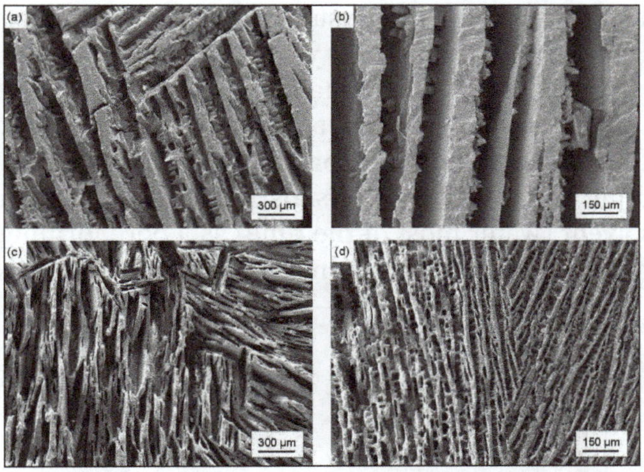

Shows the sections perpendicular (a, c) and parallel (b, d) to the direction of ice growth of samples obtained using different cooling temperature.

Equipment

- Instruments aimed to study and set up stabilized suspensions: electroacoustic technique for studying Zeta potential in function of pH and to define the optimal dispersing agent.
- Techniques for preparing stable suspensions through wet mixing.
- Freeze Dryer (Edwards) to perform the controlled freezing of aqueous based suspensions, the ice sublimation and drying of the porous structure.
- Furnaces with control of heating rates and of the atmosphere to perform thermal treatments to consolidate and assure the appropriate chemico-compositional characteristics of the porous structure.

- Instruments for microstructural investigations (scansion electronic microscopy and mercury porosimetry) and compositional detections (chemical analysis, X-ray diffraction).

Freeze Casting of Porous Ceramics

Although porosity in technical ceramics has been considered as problematic for a long time, the potentialities offered by porous ceramics are drawing considerably more attention today than just a few years ago. Cellular ceramics can be engineered to combine several advantages inherent from their architecture: they are lightweight, can have open or closed porosity making them useful as insulators or filters, can withstand high temperatures and exhibit high specific strength, in particular in compression. Typical processing methods include foam or wood replication, direct foaming or extrusion. The full potential of cellular ceramics will only be achieved once a proper control of the size, shape and amount of porosity will be available. Although the control over the structure and functional properties of cellular ceramics is continuously improving, all processing routes suffer from an inherent limitation: every processing route is intrinsically limited to a narrow range of pores characteristics. In addition, removal of the pore forming agent can be a considerable problem, and efforts have been in put in developing processing routes with environmental friendly pore forming agents, yielding techniques such as gel casting, direct foaming or recent developments with particles-stabilized wet foams. In the pursuit of such processing routes, freeze-casting has attracted considerably more focus in the last few years. The technique consists of freezing a liquid suspension (aqueous or not), followed by sublimation of the solidified phase from the solid to the gas state under reduced pressure, and subsequent sintering to consolidate and densify the walls. A porous structure is obtained, with unidirectional channels in the case of unidirectional freezing, where pores are a replica of the solvent crystals. The technique seems to be rather versatile and the use of a liquid solvent (water most of the time) as a pore forming agent is a strong asset. Freeze-casting has also been developed as a near net shape forming route, yielding dense ceramics.

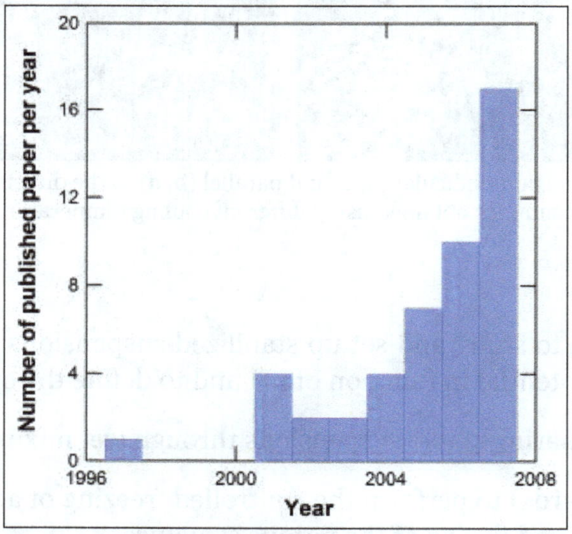

Evolution of papers published on freeze-casting of porous ceramics per year.

Processing Principles

Freeze-casting has first been developed as a near net shape forming technique, yielding dense ceramics parts with fine replicate of the mould details. Any ice crystal being converted into porosity later on in the process, introducing large size defects largely unwelcome in ceramic applications, a great deal of efforts has been put in controlling and avoiding the formation of ice crystals. Only later on was it realized that the formation and growth of ice crystals could be a substantial benefit if properly controlled, yielding porous ceramics with a very specific porosity. The early work of Fukasawa on alumina revealed the potentialities offered for porous ceramics, and a great deal of efforts has been subsequently put.

The technique consists of freezing a liquid suspension (aqueous or not), followed by sublimation of the solidified phase from the solid to the gas state under reduced pressure, and subsequent sintering to consolidate and densify the walls, leading to a porous structure with unidirectional channels in the case of unidirectional freezing, where pores are a replica of the solvent crystals.

In freeze-casting, the particles in suspension in the slurry are rejected from the moving solidification front and piled up between the growing cellular solvent crystals, in a similar way to salt and biological organisms entrapped in brine channels in sea ice. The variety of materials processed by freeze-casting suggests that the underlying principles of the technique are not strongly dependent on the materials but rely more on physical rather than chemical interactions. The phenomenon is very similar to that of unidirectional solidification of cast materials and binary alloys, in particular when powders with small (submicronic) particles size are used, with the solvent playing the role of a fugitive second phase.

The processing can be divided in four steps, and the corresponding experimental conditions will strongly depend on the chosen solvent.

1. Preparation of the slurry: This step is very similar to the preparation of slurries for conventional processing routes such as slip casting. The ceramic powder must be correctly dispersed in the liquid medium (the solvent), so that dispersant and plasticizer are often used. The temperature of the slurry must fall in the range were the solvent is liquid, room temperature in the case of water, but different temperature (60 °C and 8 °C) is necessary for respectively camphene-based and tert-butyl alcohol slurries. Moderate solid loading is used (10-40 vol%), depending of the desired amount of total porosity. The stability of the suspension must be carefully controlled to avoid any segregation phenomenon taking place in the second stage, yielding gradients of density and porosity in the final materials. This can be particularly problematic for low solid loading. Finally, the presence of a binder is necessary, to provide green strength after sublimation. Though the solvent is playing the role of the structuring agent, binder and pore forming agent, it is nevertheless removed during the sublimation stage, so that green bodies collapse in absence of an organic binder.

2. Controlled solidification of the slurry: This is the critical stage where the structure is formed and the characteristics of the future porosity are determined. During this stage, continuous crystals of solvent are formed, under certain conditions, and grow into the slurry. Ceramic particles in suspension in the slurry are rejected by the moving solidification front, concentrated and entrapped in-between the crystals. To induce this natural segregation phenomenon, the slurry is poured in a mould, which undergoes isotropic or anisotropic cooling to induce homogeneous or directional solidification. Several devices, also used to process porous polymers by freeze casting, have been designed to provide a more or less elaborated control of the solidification conditions. The solidification conditions are dictated by the initial choice of the solvent. Low temperatures (<0 °C) are

required when using water, while room temperature are sufficient when using camphene, its solidification point being around 44-48 °C. The device should also accommodate the solidification shrinkage; negative (shrinkage) in the case of camphene (-3.1%) and positive (expansion) in the case of water (+9%). The cooling conditions will largely dictate the characteristics of the growing solvent crystals and hence the final characteristics of the porosity.

3. Sublimation of the solvent: Once complete solidification of the sample is achieved, the sample is kept at conditions of low temperature and reduced pressure, conditions dictated by the physical properties of the solvent. Under these sublimation conditions, the solidified solvent is converted into the gas state. Porosity is created where the solvent crystals were, so that a green porous structure is obtained; the porosity is a direct replica of the solidified solvent structure. When using water, a conventional freeze-dryer can be used. In the case of camphene, the vapor pressure of 1.3 kPa (just below the melting temperature) is high enough to allow sublimation at room temperature, so that no specific equipment is required.

4. Sintering or densification of the green body: Once the solvent has been totally removed, the obtained green body can be sintered with conventional sintering technique. The low strength of the green body prevents any use of pressure assisted sintering. The low amount of organic binder (usually <5%) does not require the presence of a special and often problematic binder burnout process. During the sintering stage, microporosity can be removed from the ceramic walls, but the macroporosity created by the solvent crystals is retained.

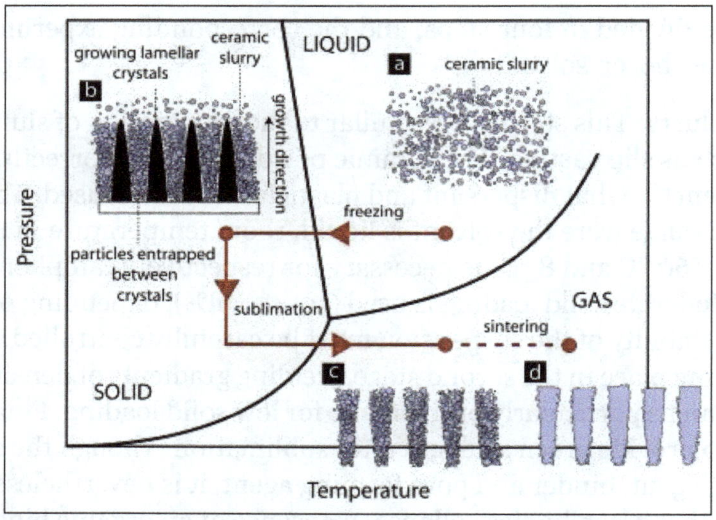

The four processing steps of freeze-casting: slurry preparation, solidification, sublimation and sintering.

Materials

A wide variety of ceramic materials have already been tested, including alumina, hydroxyapatite and tricalcium phosphate, NiO-YSZ, Ni-YSZ, yttria-stabilised zirconia, titanium dioxide, silicon nitride, PZT-PZN, mullite (from alumina gel with ultrafine silica), glass, silica, silica-alumina, clay, LSCF-CGO, MgO (from magnesium sulfate) and silicon carbide. Porous polymerceramic composites have also been processed using the technique. The variety of materials processed by the technique suggests that the underlying principles dictating the structure formation mechanisms rely on physical interactions, making the process a versatile one.

An interesting variation was reported by Yoon et al. Highly porous silicon carbide with unidirectional porosity was processed at low temperature, starting from freeze-casted green bodies of silicon carbide precursor (or a mix of powder and precursor), polycarbosilane in that case, and using camphene as a solvent. The green bodies were pyrolyzed at 1400 °C under argon, a consolidation temperature much lower than that used for the recrystallization of silicon carbide (typically 2200 °C).

Structure

Macroporosity

The porosity of the sintered materials is a replica of the original solvent crystals. A variety of pores morphology can be obtained, depending on the choice of the solvent, slurries formulation and the solidification conditions. Since the solidification is often directional, the porous channels run from the bottom to the top of the samples. Homogeneous freezing (i.e., cooling of the fingers at constant rate starting from room temperature) results in a more homogeneous ice nucleation leading to a lamellar porous architecture, with long range order, both in the parallel and perpendicular directions of the ice front. After sintering, the ceramics walls can be completely dense with no residual porosity, depending on the sintering conditions.

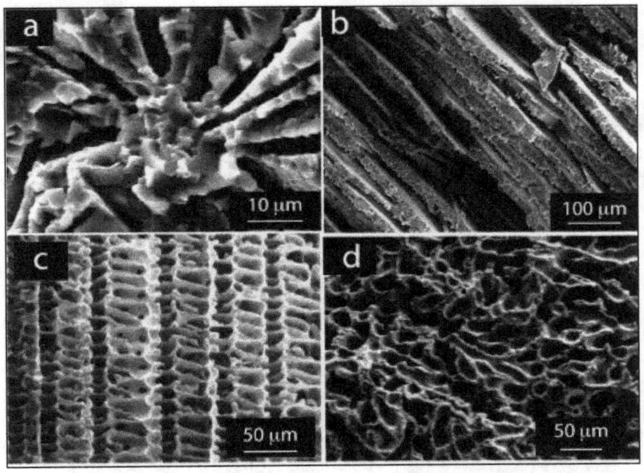

Figure indicates typical microstructures obtained by freeze-casting: (a) porous alumina using an hypoeutectic camphor/naphthalene as a solvent, (b) porous alumina using water as a solvent, (c) porous silicon carbide using polycarbosilane as a precursor and camphene as a solvent and (d) porous alumina using camphene as a solvent.

In the particular case of water being used as a solvent, the microstructure is lamellar, with lamellar channels between the ceramics walls. This particular morphology can be understood with reference to the basic crystallographic and crystal growth characteristics of ice. The ice front velocity parallel to the crystallographic c axis is 10^2 to 10^3 times lower than perpendicular to this axis. After the transition to columnar ice occurred, ice platelets with a very large anisotropy can then be formed very fast with ice growing along the a-axes, while the thickness (along the c-axis) remains small. The freezing process is easier for crystals whose c-axes are perpendicular to the temperature gradient, such that growth along the gradient can occur in the a- or b-direction. The crystals with horizontal c-axes will therefore grow at the expense of the others and continue to grow upward, in an architecture composed of long vertical lamellar crystals with horizontal c-axes. In the final structures, the direction perpendicular to the lamellae corresponds thus to the original c-axis of ice crystals.

Similar explanations can be invoked to explain the morphology of the porosity in the sintered ceramics. In the case of camphene, the solidification of liquid camphene leads to the formation of clearly defined dendrites, which are reflected in the final structures. Prismatic channels were obtained using tert-butyl alcohol as a solvent. Using other solvents will likely provide different types of morphologies.

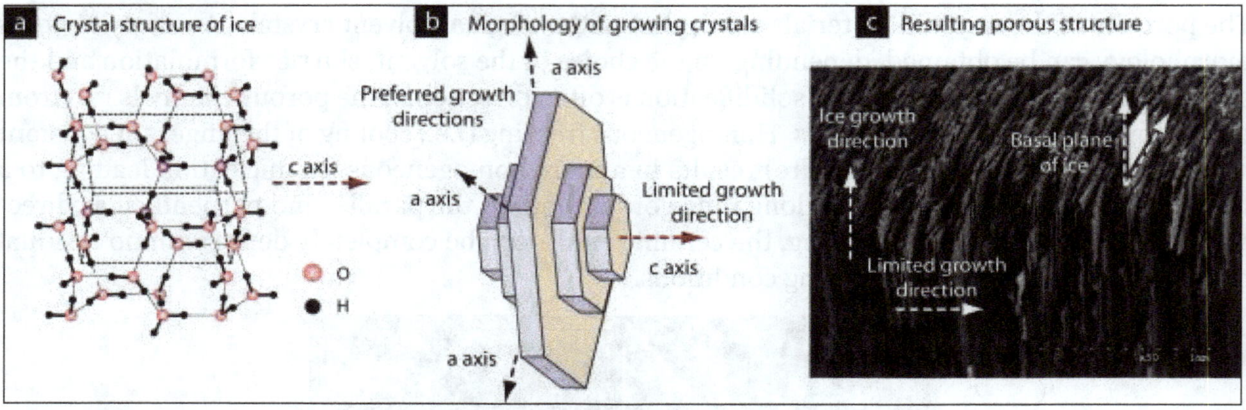

Crystal structure of ice (a) and anisotropy of crystal growth kinetics (b), leading to lamellar ice crystals. The anisotropy of the growth kinetics is reflected in the final porous structures (c) obtained after sublimation and sintering. The direction perpendicular to the ceramic platelets corresponds to the limited growth direction of ice crystals.

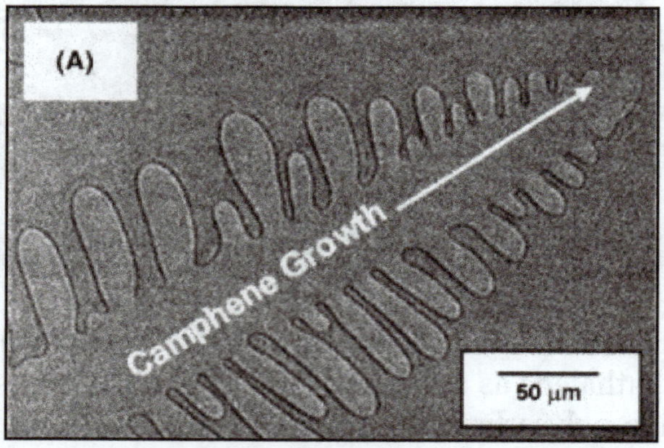

Solidified camphene dendrites, leading to the formation of a dendritic porous structure.

Orientation of Macroporosity

The pore channels can be oriented, depending on the solidification conditions. In most of the cases, the mould in which the slurry is initially poured is left with its bottom in part in contact with a cold surface. The solvent crystals are therefore solicited to grow vertically, along the direction of the imposed thermal gradient. However, different thermal gradients can be imposed, to induce a different anisotropy in the structure. A neat example of such control can be found in the study of Moon et al. Ice was stimulated to grow in the radial direction of the setup, from the inner surface of the metal cylinder to its centre region where a Teflon rod was placed. The final tubular structure exhibits a radially oriented porosity, extending from the inside towards the outside of the tube. This type of structure could ideally be used for SOFC.

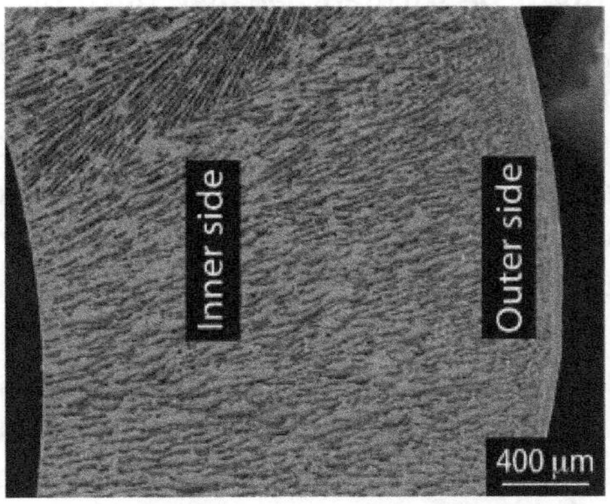

Radial porosity in a tubular structure.

Surface Roughness of Walls

The surface of the lamellae exhibits a particular topography, with dendritic-like features, running in the solidification direction. These features are homogeneous in size and distribution, but their relative size varies with the freezing conditions, the nature of the solvent, the characteristics of the starting powders and the sintering conditions (atmosphere). In the particular case of silicon nitride, the microstructure reveals the presence of elongated fibrous grains protruding from the walls. This particular feature is not related to the freezing conditions, but only appears after sintering at rather high temperatures, and is believed to result from a vapor-solid transformation taking place during sintering. A different roughness is obtained when camphene is used instead of water. In general, since the roughness is directly related to the morphology of the solvent crystal, every solvent will yield a different roughness.

Figure indicates dendritic surface of: (a) alumina using water as a solvent, (b) silicon nitride using water as a solvent after T. Fukasawa, synthesis of porous silicon nitride with unidirectionally aligned channels using freeze-drying process and hydroxyapatite using camphene as a solvent.

Ceramic Bridges

Other features of these structures are the trans-lamellar ceramic bridges, of two types. The first type just corresponds to overgrown dendrites that eventually bridge the gap between two adjacent lamellae. The second type is found in samples made from concentrated slurries. These numerous fine features with often-tortuous morphologies are locally bridging the gap between two adjacent lamellae. The morphology of these features is sometime quite different to that of the dendrites covering the ceramics lamellae, suggesting another formation mechanism. It has been proposed that they might be formed because of the specific conditions encountered during the slow freezing of highly concentrated solutions. The interaction of inert particles and a moving solidification front has been investigated for suspensions with low particles content. In such a case, the interaction between particles is not taken into account, which considerably simplifies the associated formalism. In the case of highly concentrated solutions, the particle-particle interactions cannot be ignored anymore. Eventually, it might considerably affect the pattern formation mechanisms. It has previously been shown that the particles themselves may induce morphological transitions, such as dendrites tip splitting or healing during growth before being captured. Ceramic bridges between lamellae may arise from local ice crystal tip splitting and engulfment of particle agglomerates created by particles repelled from the ice-water interface and subsequent tip healing. Depending on the magnitude of tip splitting/healing, the entrapped ceramic particles might not bridge completely the gap. The phenomenon appears to be dependent on the nature of the solvent and will largely depend on the morphology of the growing dendrites.

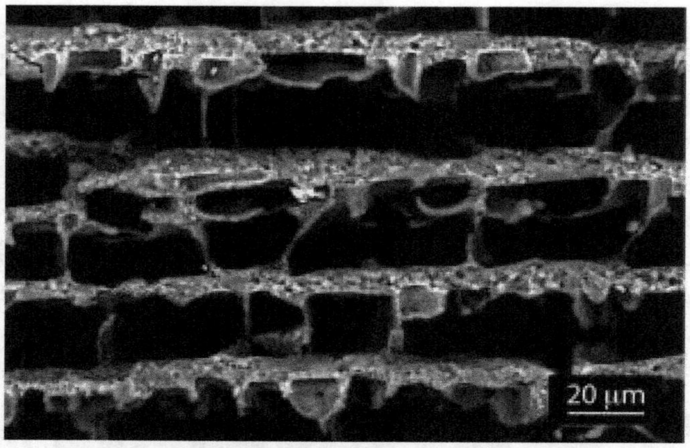

Ceramic bridges in porous lamellar alumina.

Properties of Freeze-casted Ceramics

Few properties of freeze-casted ceramics have been measured so far, as most of the attention was paid to the control of the structure and the processing conditions. Preliminary reports can nonetheless be found on one of the most critical property of cellular ceramics: the compressive strength. The strength is of particular importance in the case of intrinsically weak ceramics such as calcium phosphate, being considered as potential candidates for bone replacement applications. Dramatic improvements of compressive strength of hydroxyapatite (HAP) were reported, using water as a solvent (i.e., lamellar architecture). Although for high porosity content (typically >60 vol%) the strength (16 MPa) is comparable to that reported in the literature, it increases rapidly when the porosity decreases, reaching 65 MPa at 56% porosity and 145 MPa at 47% porosity. Values obtained for these samples are well above those reported so far in the literature. In fact, the strength of the porous lamellar HAP is close to that of compact bone. Such high values allow considering the potential of these materials for some load-bearing applications. Porous hydroxyapatite was also processed by freeze-casting using camphene as a solvent, resulting in very different pores morphologies. Interestingly, the trend followed by the compressive strength seems to be highly dependent on the morphology of the pores. With camphene, the pores are dendritic and the structure somewhat close to usual cellular ceramics. With water, the pores are lamellar, and the compressive strength reaches much greater values, in particular for lower porosity values, with dramatic improvements observed. A possible explanation could be the strong anisotropy of the structure in the loading direction and the presence of inorganic bridges between the ceramic layers, which might prevent Euler buckling of the ceramic layers. Further investigations are nonetheless needed to rationalize these observations and confirm these trends.

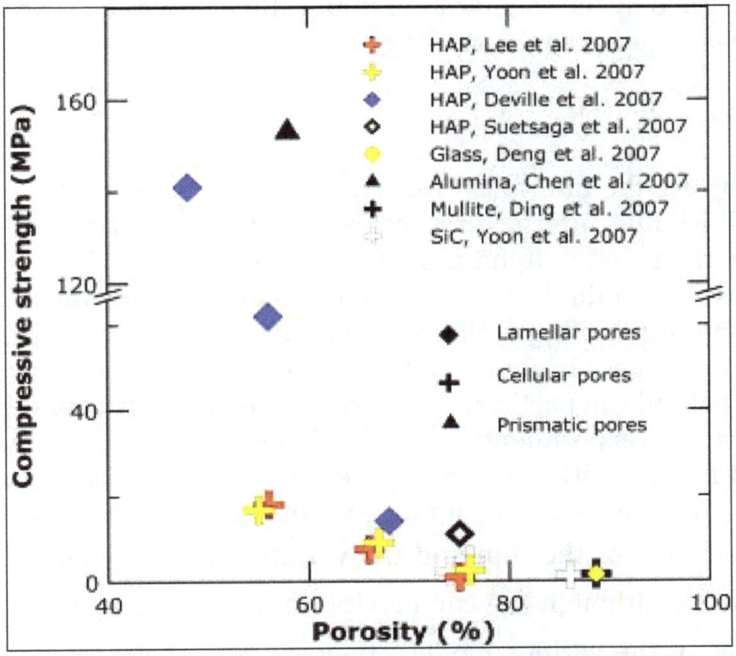

Compressive strength vs. total porosity. The compressive strength seems to be highly dependent of the morphology of the pores, with much higher values measured for freeze-casted ceramics with lamellar and prismatic morphologies.

Formation and Control of the Structure

Parameters affecting the final structure can be found in all the processing stages: slurry formulation and preparation (including the characteristics and properties of the starting powders), solidification and sintering. A few parameters are very specific and related to the core of the process: the freezing of the slurry. The solidification stage is the most critical with regards of the final porous structure; most of the features of the porosity will be created during this stage.

The formation of regular patterns is a common feature of many solidification processes, such as eutectic growth or unidirectional solidification of two-phase systems. Control of the regularity and size of the patterns is often a key issue with regards to the final properties of the materials. Many of the features of the freeze-casted porous ceramics can be understood by applying generic principles of solidification processes. In particular, the physics of water freezing has drawn the attention of scientists for a long time, for implications in fields as diverse as cryopreservation, cleaning of pollutants or even polar ice formation.

Formation of the Structure

Interaction between the Solidification front and the Ceramic Particles

In order to obtain ceramic samples with a porous structure, two requirements must be satisfied:

1. The ceramic particles in suspension in the slurry must be rejected from the advancing solidification front and entrapped between the growing ice crystals. This aspect can be understood considering the interaction between the solvent solidification front and the particles in suspensions. A simple thermodynamic criterion can be used in a first approach. The thermodynamic condition for a particle to be rejected by the solidification front is that there is an overall increase in surface energy if the particle is engulfed by the solid, i.e.,

$$\Delta\sigma = \sigma_{sp} - \left(\sigma_{lp} + \sigma_{sl}\right) > 0$$

where, σ_{sp}, σ_{lp} and σ_{sl} are the interfacial free energies associated with the solid-particle, liquid-particle and solid-liquid interface respectively. When this criterion is satisfied and the particles rejected by the front, a liquid film should exist between the solidification front and the particle in order to maintain the transport of molecules towards the growing crystal. When the velocity of the front increases, the thickness of the film decreases. There is a critical velocity, v_c, for which this thickness is not enough to allow the necessary flow of molecules to keep the crystal growing behind the particle, that becomes then encapsulated by the solid. A large amount of theoretical and experimental work has been addressing this problem, and several expressions derived for the critical velocity. A few examples can be found in the references. The main physical parameter to be taken into account includes the viscosity of the liquid, the particle size, the thickness of the film and the variation of free energy defined in equation $\Delta\sigma = \sigma_{sp} - \left(\sigma_{lp} + \sigma_{sl}\right) > 0$. Although the complexity of the system leaves enough room for discussing the relevance of the various models, they are nonetheless useful to understand the influence of the physical parameters on the behavior of the systems and the morphology of the materials resulting from these interactions.

2. The ice front must have a non-planar morphology. Indeed, if the front is planar and the particles rejected, all the particles are collected on one side of the sample once solidification is achieved. This effect is being used in the purification of pollutants. However, to form porous structures, particles redistribution must occur; the particles must be rejected from the solidification front and collected between the arms of the solidification front. The morphology of the front will then dictate the architecture of the final materials. At the very beginning of solidification, the interface (the front) is planar, and must somehow undergo a transition towards an irregular morphology, that is – cellular, lamellar or even more complex dendritic morphologies. This transition can be triggered by different mechanisms. One mechanism is the inherent thermodynamic instability of the interface, also known as Mullins-Sekerka instability. The development of the instability is based on supercooling effects, building up with solute rejection ahead of the interface. Another mechanism is related to the presence of the particles. In that case, the instability is due to the reversal of the thermal gradient in the liquid ahead of the interface and behind the particle. It is not clear at this point which of these mechanisms is dominating.

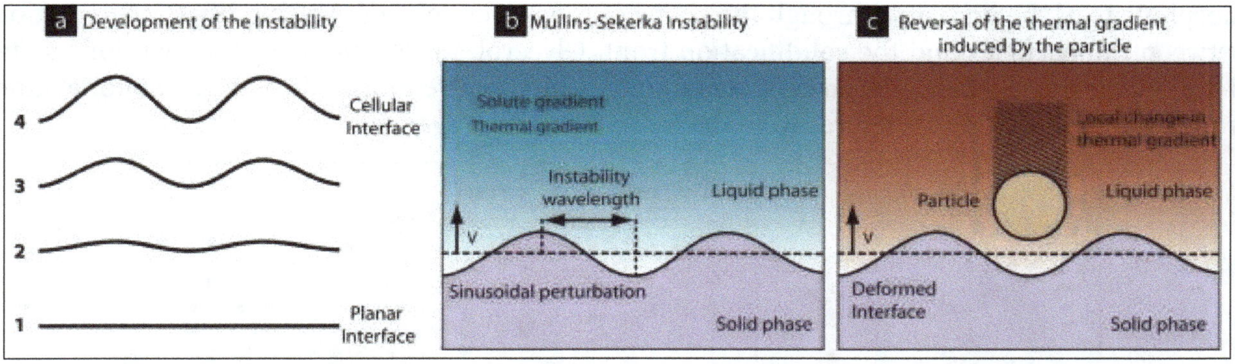

Destabilization of the interface and possible mechanisms triggering this morphological transition.

Influence of Processing and Physical Parameters

1. Solvent: The choice of the solvent will be crucial both in regards of the processing conditions and on the characteristics of the structure that is desired. If freeze-casting was originally developed using water as a solvent, alternative solutions are considered today. Requirements for such alternative includes solidification temperature, viscosity of the liquid, limited volume change associated to the solidification, high vapor pressure in solid state to allow sublimation under reasonable conditions of temperature and pressure, and, of course, environmental issues and price considerations. Among the other potential candidates, preliminary investigations with polymer systems have revealed liquid carbon dioxide (CO_2) as a suitable solvent. The structures obtained using carbon dioxide is somewhat similar to those obtained with camphene, with a complex dendritic structure.

2. Ceramic powder: The properties of the starting powder might have a major effect on the characteristics of the final materials, although little results have been reported so far. The core of the process being based on the interaction between the particles and the solidification front, a number of parameters are susceptible to modify these interactions, including the size of the particles, their size distribution, their shape, surface roughness, and surface tension. The only influence reported so far is that of the particle size, affecting the relationship between the solidification kinetics and the structure wavelength. Further investigations are needed to clarify these effects.

3. Slurry formulation: As in any ceramic process, the formulation of the slurry must be carefully optimized. To ensure a homogeneous structure in the sintered materials, any segregation effects must be avoided. Slurries must be stable during the entire duration of the freezing stage. The microstructure can also be modified by varying the concentration of the starting slurry. Since the solvent initially present in the slurry is converted first into solid that is later eliminated to form the porosity, the pore content can be adjusted by tuning the slurry characteristics. The final porosity of the material is directly related to the volume of solvent in the suspension. From the data plotted in figure it appears that a wide range of porosity content can be achieved through freeze-casting, approximately from 25 to 90%. The reasons for these limits are described in the final part of this review. The formulation has also been modified through additives, like polystyrene or glycerol, to achieve certain effects. These additives are likely to affect the viscosity, surface tension and modify the super cooling effects. Additives can be desirable to modify the

morphology of the porosity (through the shape of the solvent crystals) or modify the interaction between the particles and the solidification front. Glycerol, for example, well-known for its anti-freeze effects in other applications, was found to disrupt the morphology of the solidification front, and the resulting structures were dense.

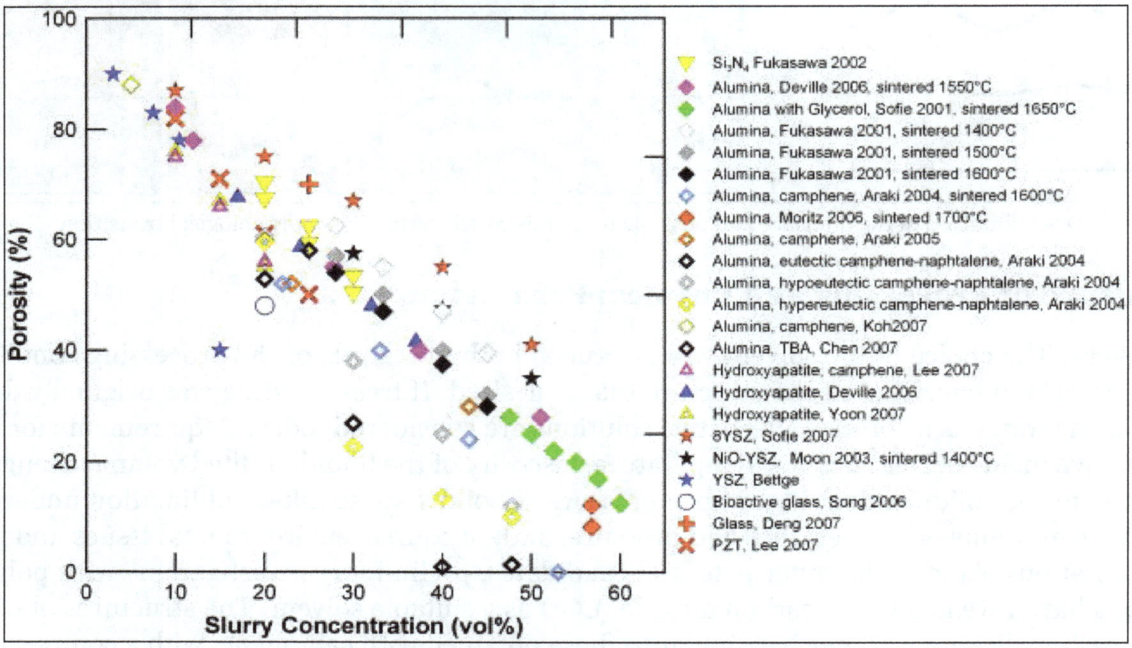

Total porosity vs. slurry solid loading.

4. Solidification conditions: The directionality of the solidification is critical in regards of the desired directionality of the porosity. When the slurries are solidified without any temperature gradient applied, the crystals can nucleate at any place and have no preferred growth direction. This results in structures with a random orientation of the porosity. However, when the experimental setup allows imposing a defined temperature gradient, the solvent crystals are forced to grow along the temperature gradient. Provided the temperature field is carefully controlled, crystals and hence resulting pore channels can run through the entire samples, reaching dimensions of a few centimeters. The tortuosity of the porosity in such cases is close to 1. Besides the directionality, the nucleation conditions can be important. If the slurry is partially quenched, i.e., poured over a cold finger maintained at a constant and negative temperature, the initial freezing is not steady. Although lamellae and channels are observed all over the sample, their orientation over the cross-section parallel to the ice front is completely random. Colonies of locally aligned pores are observed, but no long-range order is found. Homogeneous freezing (i.e., cooling of the fingers at constant rate starting from room temperature) results in a more homogeneous ice nucleation leading to an oriented and continuous lamellar porous architecture, with long range order, both in the parallel and perpendicular directions of the ice front. Solidification kinetics was also found to have a dramatic influence on the structure. When the freezing kinetics is increased, i.e., the solidification front speed increases, the width of the channels and of the lamellae is drastically affected. The faster the freezing rate, the finer is the microstructure. The empirical dependence between the wavelength λ (or the lamellae thickness) and the speed of the ice front in the direction parallel to the temperature gradient (v) can be described with a simple power law ($\lambda = A.v^{-n}$).

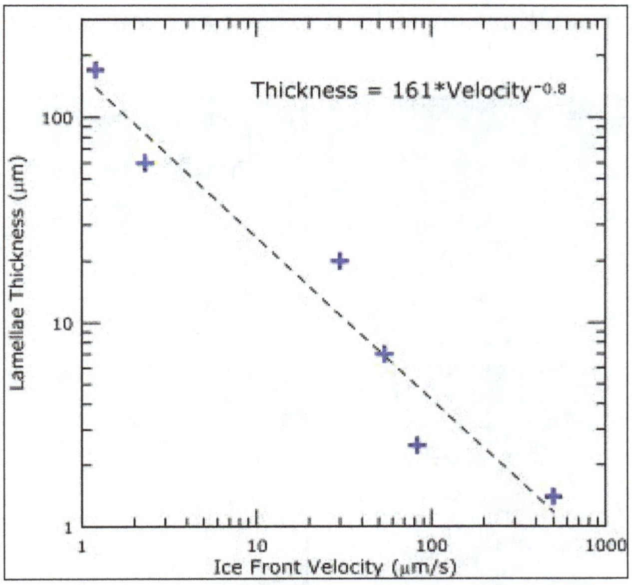

Lamellae thickness vs. ice front velocity, porous lamellar alumina.

5. Sintering: As with any other ceramic process, the sintering stage can be optimized to control the porosity/density of the final piece. Besides the densification of the walls, the effect of sintering on the structure and macroporosity characteristics has not been tackled yet, and the mechanisms controlling the shrinkage have not been identified. The only modification reported is the formation of elongated fibrous grains during the sintering of freeze-casted porous silicon nitride, which was believed to be due to some vapor-solid phase transformation.

Limits

Solid Content in Slurry and Green Body Density

If the particles are rejected by the solidification front, the particle concentration in the remaining melt areas will become larger. Dendrites of the solidified solvent grow into the liquid, pushing the ceramic particles into the interdendritic spaces. Eventually, the particle redistribution ceases, and the solid/liquid interface moves into the interparticles spaces. This phenomenon has been called a "breakthrough" as the solid/liquid interface breaks into the suspension. A simple yet efficient model has been developed by Shanti et al., taking into account the capillary force pushing the particles with the interface and the countering force, i.e., the resultant of the osmotic pressure of the suspension. The particle volume fraction at breakthrough was found to be determined by the maximum packing of particles at the point of jamming, modified by a small term dependent upon particle size and surface tension. The predictions agreed with observation for the alumina–camphene system within about 4%, which is quite satisfying considering the simplicity of the model used. As a consequence, the density of the green body after sublimation is always the same, providing the formulation of the slurry is kept constant. Little influence of the nature of the solvent and characteristics of the particles was found. Another consequence of this analysis is the existence of a maximum solid content in the formulation of the slurry. If the solid content already exceeds the breakthrough concentration, the solid/liquid interface moves into the interparticles space before any particle redistribution can occur. Particles distribution in the frozen samples is homogeneous, and the green body obtained after sublimation will not exhibit any continuous porosity resulting from particles-free dendrites.

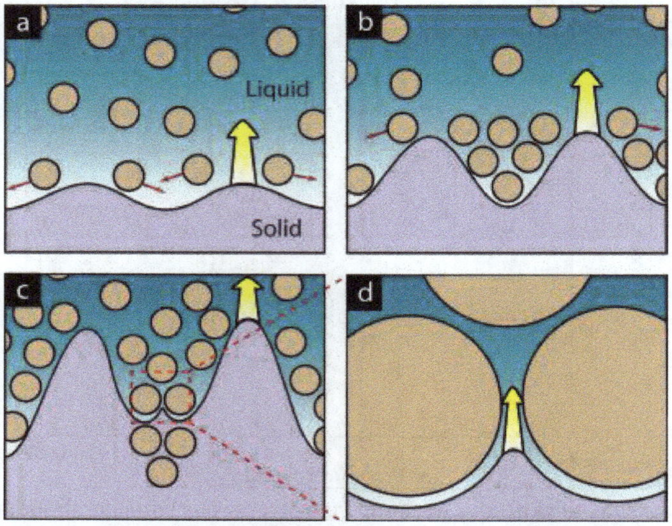

Particles redistribution during solidification. Particles are concentrated between the solidified dendrites (a-b) until the capillary pressure exceeds the osmotic pressure (c), at which point the solid/liquid interface moves into the interparticles space (d).

Directionality of the Porosity

The directionality of the porosity can be controlled by the solidification conditions, unidirectional solidification yielding similarly oriented structures. Refinements have nonetheless be developed, such as solidification in a radial direction, to process ceramic tubes with a radial porosity, a feature making them useful as filters. Inducing a complex orientation of the porosity will directly depend on the ability to properly and homogeneously control the direction of propagation of the solidification front, becoming thus an experimental setup issue. Inspiration can also be taken from investigations of porous polymers obtained through the same approach, for example where complex interpenetrating materials were obtained. This was done by freezing in the same direction as the original aligned channels or perpendicular to it. Rather unique structures were obtained with grid-like morphologies. Some of these concepts could likely be extended to porous ceramics.

Walls Thickness, Pore Size and Solidification Kinetics

A number of studies have revealed the relationships between the solidification kinetics (i.e., the solidification front speed) and the size of the porosity (or the structure) of the freeze-casted ceramics. Such relationships are commonly found in processing techniques based on directional solidification, including metallic and polymeric materials. Criteria were developed to predict the dendritic spacing as a function of the solidification kinetics, and several experimental parameters have been identified. The faster the freezing rate, the finer the resulting structure, as shown before. Limits are nevertheless encountered in the case of porous ceramics, due to the underlying phenomena: the solidification of the solvent and the interaction between the solidification front and the ceramic particles. Preliminary results can be found in reference, although a lot of work is still necessary to investigate these limits. One limit is related to the critical velocity for entrapment. Above this critical velocity, particles are entrapped by the moving interface and porosity is disappearing from the final structure. Upgrading the experimental setup to increase the freezing kinetics will be helpless with this issue beyond a certain point, and obtaining cooling rate in the range of these obtained during the solidification of metallic alloys will be useless. Several parameters seem nonetheless to affect this critical velocity, the most important being probably the solid content in the slurry, the particle size and the properties of the solvent. These limits are schematically summarized in figure.

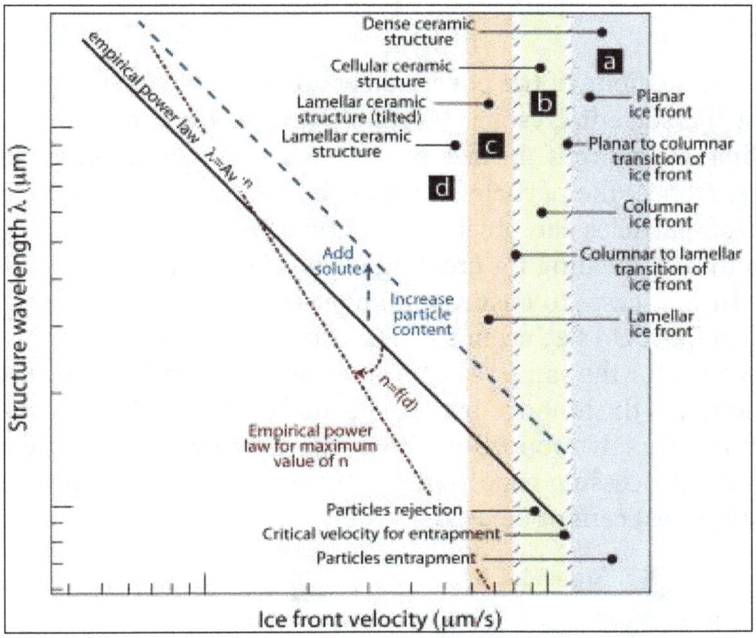

The above figure indicates strategies and limits for controlling the structure: schematic plot of wavelength vs. ice front velocity. The exponent n of the empirical law is dependent on the particle size d, though the function n = f(d) is not monotonic; an optimum of d is encountered where the exponent n is maximum. Very fast cooling rates, region (a), will result in the ice front trapping the particles and the formation of a dense material. When the velocity is decreased below the critical value for particle entrapment, v_c, the particles are expelled from the growing ice but if the speeds are fast enough the ice will grow with a columnar microstructure, region (b). Slower velocities will result in the formation of lamellar ice. However, if the velocity is still fast or, equivalently, the gradient in temperature small enough, the balance between the preferential growth direction and the gradient direction will result in the growth of lamellae tilted with respect to the later, region (c). As the velocity decreases (or the gradient increases) the lamellae will align with the direction of the temperature gradient, region (d).

Pores Morphology

The morphology and characteristics of the pores are influenced by several independent or interconnected parameters, as shown previously. Any modification of these parameters will have a direct repercussion on the structure of the materials. Three main parameters have nevertheless been identified as being critical with regards to this issue:

1. Nature of the solvent: This one is probably the most critical one. The crystallographic properties of the solvent in its solid state will define the main appearance of the structure, i.e., lamellar with water, dendritic with camphene or prismatic with tert-butyl alcohol. Obtaining radically different

morphologies will imply a new choice of solvent with different properties, and will be restricted by the requirements exposed previously.

2. Freezing conditions: Both the kinetics and the directionality of the freezing conditions can be used to modify the pores morphology. For the reasons previously exposed, the freezing conditions will affect both the pore size and the pore morphology, so that these two parameters are usually interdependent.

3. Particle size: The homogeneity of the porous structure is lost when the particle size become too similar to the size of the solvent crystals. Morphological features of the crystals cannot be well replicated into the final structure if the size of the particle is in the range of order of that of the interdendritic spaces. Hence, the particle size must be kept below the wavelength of the desired structure. Micron-sized particles can obviously not be used to create nanometer-sized structure. Freeze-casting seem to be appealing for creating porous structures with a pore size smaller than a hundred microns. In that case, to ensure a homogeneous and well defined structure, powders with submicronic mean particle size seem to be desirable. Distribution of the particle size should also be taken into account for the same reasons, and the presence of large particle or particles agglomerates is detrimental to the homogeneity of the final structure. Finally, the critical velocity for particles entrapment is inversely proportional to the particle size. Using larger particles will lower this critical velocity and processing conditions might fall in the range were fast freezing kinetics (desirable to get small pores) cannot be used.

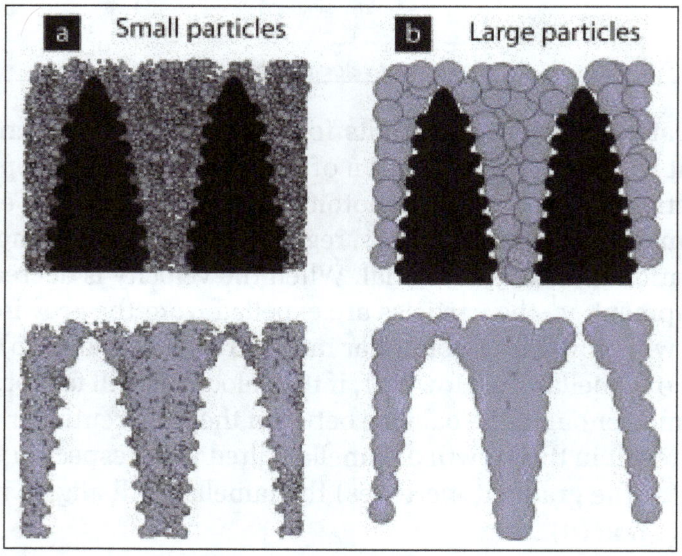

Influence of particle size on pores morphology. (a) Shows details of the crystals are well replicated in the case of small particles; (b) such details are lost with large particles.

Processing Time and Dimensions of the Pieces

Two stages of the process are critical in regards of the total processing time and the dimensions of the final pieces: the freezing stage and the sublimation stage.

1. Freezing time: The freezing time required will depend on two parameters: the desired structure wavelength (i.e., pore size or wall thickness) and the dimensions of the final piece. Regarding the

structure wavelength, freeze-casting seem to offer unique opportunities for dimensions below a hundred microns. Within this range, the freezing kinetics required is rapid enough so that the processing times remain reasonable (a few minutes for pieces of a few centimeters). Reaching smaller pores will imply faster freezing, so that the freezing time is even further reduced. The desired dimensions of the pieces must also be taken into account. Most of the investigations reported so far were dealing with sample's thickness in the range 1-3 cm, i.e., at the lab scale. Freezing time will more or less linearly increase with the thickness of the sample, but maintaining the homogeneity of the temperature field and the freezing kinetics (which control the homogeneity of the porosity) will become problematic for pieces of larger dimensions. This will imply both a better control of the formulation to ensure its stability during the solidification and improved experimental setup with optimal temperature control.

2. Sublimation: The time necessary for sublimating the solvent is directly dependent on the dimensions of the pieces to be treated. While kept in the centimeter range, sublimation can occur overnight. This stage must nevertheless be carefully controlled to avoid any formation of defects (cracks, distortion) during the solid-vapor phase transition, and such control might become problematic for larger pieces. Technical solutions are already available to treat pieces of larger dimensions.

Applications of Porous Freeze-casted Ceramics

The potentialities offered by the unique structures obtained by freeze-casting have drawn attention for a number of very diverse applications, although very few figures of functional properties have been reported so far. The applications considered so far belongs to two main groups:

1. Biomaterials: The structure of the porous ceramics obtained by freeze-casting being very similar to that of natural biomaterials such as nacre or bone, biomaterials has become a field with potentially important applications of this technology, with particular attention to tissue engineering. Preliminary reports of dramatic improvements of the compressive strength of freeze casted hydroxyapatite seem to confirm that potential, and the assessment of the biological response of these materials is under way in various laboratories.

2. Materials for chemical processes and energy sources: Including SOFC, electrodes, catalysts, sensors, filtration/separation devices and photocatalysis for air or water purification. The unique morphology of the porosity is of interest for these applications, where open structures with good connectivity and low tortuosity (close to 1, ensuring optimal access of the fluids to the reaction sites), large surface area (large number of electrochemical reaction sites), dense ceramic walls (good conductivity) and good mechanical properties are required. The directionality of the porosity can be used to optimize the species transport properties while the dense walls will ensure optimal conducting properties and mechanical reliability. The versatility of the process allows considering a wide range of technical ceramics or composites (e.g., NiO-YSZ), and the typical dimensions of the samples processed so far appear to be suitable for most of these applications.

A recent report of freeze-casted piezoelectrics can also be found, with high hydrostatic figures of merit. Such materials could find applications as actuators, transducers and in particular low frequency hydrophones.

References

- An-overview-of-the-firing-process: thesprucecrafts.com, Retrieved 21 March , 2019
- Extrusion-basics: ceramicindustry.com, Retrieved 12 April, 2019
- How-to-slipcast-ceramics: thesprucecrafts.com, Retrieved 23 June, 2019
- Dry: capa-ceramic.com, Retrieved 3 August, 2019
- Ceramic-mold-casting: thelibraryofmanufacturing.com, Retrieved 19 May, 2019
- Tape-casting, materials-science: sciencedirect.com, Retrieved 28 April, 2019
- Freeze-casting, 109-forming-techniques, processes, resources: istec.cnr.it, Retrieved 8 August, 2019

5

Ceramics Manufacturing

Ceramics are used in numerous manufacturing processes such as coating, machining and joining. Ceramic materials are also used to make cutting tools. The diverse applications of ceramics in these manufacturing processes as well as the hardness testing of ceramics have been thoroughly discussed in this chapter.

Ceramic Products Manufacturing

Ceramic products that are made from highly refined natural or synthetic compositions and designed to have special properties are referred to as advanced ceramics. Advanced ceramics can be classified according to application as electrical, magnetic, optical, chemical, thermal, mechanical, biological, and nuclear.

Most ceramic products are clay-based and are made from single clay or one or more clays mixed with mineral modifiers such as quartz and feldspar. The types of commercial clays used for ceramics are primarily kaolin and ball clay.

Process Description

Figure presents a general process flow diagram for ceramic products manufacturing. The basic steps include raw material procurement, beneficiation, mixing, forming, green machining, drying, presinter thermal processing, glazing, firing, final processing, and packaging.

Raw Material Procurement

To begin the process, raw materials are transported and stored at the manufacturing facility. The raw materials used in the manufacture of ceramics range from relatively impure clay materials mined from natural deposits to ultrahigh purity powders prepared by chemical synthesis. Naturally occurring raw materials used to manufacture ceramics include silica, sand, quartz, flint, silicates, and aluminosilicates (e. g., clays and feldspar).

Beneficiation

The next step in the process is beneficiation. Although chemically synthesized ceramic powders also require some beneficiation, the focus of this discussion is on the processes for beneficiating naturally occurring raw materials. The basic beneficiation processes include comminution, purification, sizing, classification, calcining, liquid dispersion, and granulation. Naturally occurring raw materials often undergo some beneficiation at the mining site or at an intermediate processing facility prior to being transported to the ceramic manufacturing facility.

Comminution entails reducing the particle size of the raw material by crushing, grinding, and milling or fine grinding. The purpose of comminution is to liberate impurities, break up aggregates, modify particle morphology and size distribution, facilitate mixing and forming, and produce a more reactive material for firing. Primary crushing generally reduces material up to 0.3 meter (m) (1 foot) in diameter down to 1 centimeter (cm) (0.40 inch) in diameter. Secondary crushing reduces particle size down to approximately 1 millimeter (mm) (0.04 in) in diameter. Fine grinding or milling reduces the particle size down to as low as 1.0 micrometer (μm) (4×10^{-5} in) in diameter. Ball mills are the most commonly used piece of equipment for milling. However, vibratory mills, attrition mills, and fluid energy mills also are used. Crushing and grinding typically are dry processes; milling may be a wet or dry process. In wet milling, water or alcohol commonly is used as the milling liquid.

Several procedures are used to purify the ceramic material. Water soluble impurities can be removed by washing with deionized or distilled water and filtering, and organic solvents may be used for removing water-insoluble impurities. Acid leaching sometimes is employed to remove metal contaminants. Magnetic separation is used to extract magnetic impurities from either dry powders or wet slurries. Froth flotation also is used to separate undesirable materials.

Sizing and classification separate the material into size ranges. Sizing is most often accomplished using fixed or vibrating screens. Dry screening can be used to sizes down to 44 μm (0.0017 in, 325 mesh). Dry forced-air sieving and sonic sizing can be used to size dry powders down to 37 μm (0.0015 in, 400 mesh), and wet sieving can be used for particles down to 25 μm (0.00098 in, 500 mesh). Air classifiers generally are effective in the range of 420 μm to 37 μm (0.017 to 0.0015 in, 40 to 400 mesh). However, special air classifiers are available for isolating particles down to 10 μm (0.00039 in).

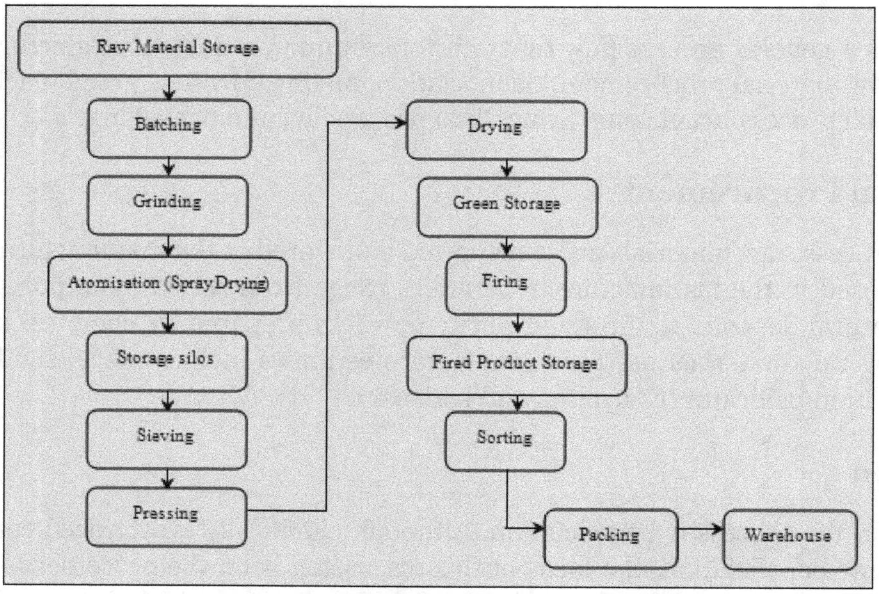

Process flow diagram for ceramic products manufacturing.

Calcining consists of heating a ceramic material to a temperature well below its melting point to liberate undesirable gases or other material and to bring about structural transformation to produce the desired composition and phase product. Calcining typically is carried out in rotary calciners, heated fluidized beds, or by heating a static bed of ceramic powder in a refractory crucible.

Liquid dispersion of ceramic powders sometimes is used to make slurries. Slurry processing facilitates mixing and minimizes particle agglomeration. The primary disadvantage of slurry processing is that the liquid must be removed prior to firing the ceramic.

Dry powders often are granulated to improve flow, handling, packing, and compaction. Granulation is accomplished by direct mixing, which consists of introducing a binder solution during powder mixing, or by spray drying. Spray dryers generally are gas-fired and operate at temperatures of 110° to 130 °C (230° to 270 °F).

Mixing

The purpose of mixing or blunging is to combine the constituents of a ceramic powder to produce a more chemically and physically homogenous material for forming. Pug mills often are used for mixing ceramic materials. Several processing aids may be added to the ceramic mix during the mixing stage. Binders and plasticizers are used in dry powder and plastic forming; in slurry processing, deflocculants, surfactants, and antifoaming agents are added to improve processing. Liquids also are added in plastic and slurry processing.

Binders are polymers or colloids that are used to impart strength to green or unfired ceramic bodies. For dry forming and extrusion, binders amount to 3 percent by weight of the ceramic mixture. Plasticizers and lubricants are used with some types of binders. Plasticizers increase the flexibility of the ceramic mix. Lubricants lower frictional forces between particles and reduce wear on equipment. Water is the most commonly used liquid in plastic and slurry processing. Organic liquids such as alcohols may also be used in some cases. Deflocculants also are used in slurry processing to improve dispersion and dispersion stability. Surfactants are used in slurry processing to aid dispersion, and antifoams are used to remove trapped gas bubbles from the slurry.

Forming

In the forming step, dry powders, plastic bodies, pastes, or slurries are consolidated and molded to produce a cohesive body of the desired shape and size. Dry forming consists of the simultaneous compacting and shaping of dry ceramic powders in a rigid die or flexible mold. Dry forming can be accomplished by dry pressing, isostatic compaction, and vibratory compaction.

Plastic molding is accomplished by extrusion, jiggering, or powder injection molding. Extrusion is used in manufacturing structural clay products and some refractory products. Jiggering is widely used in the manufacture of small, simple, axially symmetrical whiteware ceramic such as cookware, fine china, and electrical porcelain. Powder injection molding is used for making small complex shapes.

Paste forming consists of applying a thick film of ceramic paste on a substrate. Ceramic pastes are used for decorating ceramic tableware, and forming capacitors and dielectric layers on rigid substrates for microelectronics.

Slurry forming of ceramics generally is accomplished using slip casting, gelcasting, or tape casting. In slip casting, ceramic slurry, which has a moisture content of 20 to 35 percent, is poured into a porous mold. Capillary suction of the mold draws the liquid from the mold, thereby consolidating the cast ceramic material. After a fixed time the excess slurry is drained, and the cast is dried. Slip casting is widely used in the manufacture of sinks and other sanitaryware, figurines, porous thermal insulation, fine china, and structural ceramics with complex shapes. Gelcasting uses in situ polymerization of organic monomers to produce a gel that binds ceramic particles together into complex shapes such as turbine rotors. Tape casting consists of forming a thin film of ceramic slurry of controlled thickness onto a support surface using a knife edge. Tape casting is used to produce thin ceramic sheets or tape, which can be cut and stacked to form multilayer ceramics for capacitors and dielectric insulator substrates.

Green Machining

After forming, the ceramic shape often is machined to eliminate rough surfaces and seams or to modify the shape. The methods used to machine green ceramics include surface grinding to smooth surfaces, blanking and punching to cut the shape and create holes or cavities, and laminating for multilayer ceramics.

Drying

After forming, ceramics must be dried. Drying must be carefully controlled to strike a balance between minimizing drying time and avoiding differential shrinkage, warping, and distortion. The most commonly used method of drying ceramics is by convection, in which heated air is circulated around the ceramics. Air drying often is performed in tunnel kilns, which typically use heat recovered from the cooling zone of the kiln. Periodic kilns or dryers operating in batch mode also are used. Convection drying also is carried out in divided tunnel dryers, which include separate sections with independent temperature and humidity controls. An alternative to air drying is radiation drying in which microwave or infrared radiation is used to enhance drying.

Presinter Thermal Processing

Prior to firing, ceramics often are heat-treated at temperatures well below firing temperatures. The purpose of this thermal processing is to provide additional drying, to vaporize or decompose organic additives and other impurities, and to remove residual, crystalline, and chemically bound water. Presinter thermal processing can be applied as a separate step, which is referred to as bisque firing, or by gradually raising and holding the temperature in several stages.

Glazing

For traditional ceramics, glaze coatings often are applied to dried or bisque-fired ceramic ware prior to sintering. Glazes consist primarily of oxides and can be classified as raw glazes or frit glazes. In raw glazes, the oxides are in the form of minerals or compounds that melt readily and act as solvents for the other ingredients. Some of the more commonly used raw materials for glazes are quartz, feldspars, carbonates, borates, and zircon. A frit is a prereacted glass.

To prepare glazes, the raw materials are ground in a ball mill or attrition mill. Glazes generally are applied by spraying or dipping. Depending on their constituents, glazes mature at temperatures of 600° to 1500 °C (1110° to 2730 °F).

Firing

Firing is the process by which ceramics are thermally consolidated into a dense, cohesive body comprised of fine, uniform grains. This process also is referred to as sintering or densification. In general: (1) ceramics with fine particle size fire quickly and require lower firing temperatures; (2) dense unfired ceramics fire quickly and remain dense after firing with lower shrinkage; and (3) irregular shaped ceramics fire quickly. Other material properties that affect firing include material surface energy, diffusion coefficients, fluid viscosity, and bond strength.

Parameters that affect firing include firing temperature, time, pressure, and atmosphere. A short firing time results in a product that is porous and has a low density; a short to intermediate firing time results in fine-grained (i. e., having particles not larger than 0.2 millimeters), high-strength products; and long firing times result in a coarse-grained products that are more creep resistant. Applying pressure decreases firing time and makes it possible to fire materials that are difficult to fire using conventional methods. Oxidizing or inert atmospheres are used to fire oxide ceramics to avoid reducing transition metals and degrading the finish of the product.

In addition to conventional firing, other methods used include pressure firing, hot forging, plasma firing, microwave firing, and infrared firing.

Conventional firing is accomplished by heating the green ceramic to approximately two-thirds of the melting point of the material at ambient pressure and holding it for a specified time in a periodic or tunnel kiln. Periodic kilns are heated and cooled according to prescribed schedules. The heat for periodic kilns generally is provided by electrical element or by firing with gas or oil.

Tunnel kilns generally have separate zones for cooling, firing, and preheating or drying. The kilns may be designed so that: (1) the air heated in the cooling zone moves into the firing zone and the combustion gases in the firing zone are conveyed to the preheat/drying zone then exhausted or (2) the air heated in the cooling zone is conveyed to the preheat/drying zone and the firing zone gases are exhausted separately. The most commonly used tunnel kiln design is the roller hearth (roller) kiln. In conventional firing, tunnel kilns generally are fired with gas, oil, coal, or wood. Following firing and cooling, ceramics are sometimes refired after the application of decals, paint, or ink.

Advanced ceramics often are fired in electric resistance-heated furnaces with controlled atmospheres. For some products, separate furnaces may be needed to eliminate organic lubricants and binders prior to firing.

Ceramic products also are manufactured by pressure firing, which is similar to the forming process of dry pressing except that the pressing is conducted at the firing temperature. Because of its higher costs, pressure firing is usually reserved for manufacturing ceramics that are difficult to fire to high density by conventional firing.

Final Processing

Following firing, some ceramic products are processed further to enhance their characteristics or to meet dimensional tolerances. Ceramics can be machined by abrasive grinding, chemical polishing, electrical discharge machining, or laser machining. Annealing at high temperature, followed by gradual cooling can relieve internal stresses within the ceramic and surface stresses due to machining. In addition, surface coatings are applied to many fired ceramics. Surface coatings are applied to traditional clay ceramics to create a stronger, impermeable surface and for decoration. Coatings also may be applied to improve strength, and resistance to abrasion and corrosion. Coatings can be applied dry, as slurries, by spraying, or by vapor deposition.

Emissions and Controls

The primary pollutants associated with raw material beneficiation are particulate matter (PM) and PM less than 10 μm in aerodynamic diameter (PM-10). Filterable PM and PM-10 are emitted from comminution, sizing, classifying, handling, transfer, and storage. In addition, raw material calciners emit filterable and condensible PM, which may include metals and other inorganic pollutants. Calciners also emit products of combustion such as nitrogen oxides (NO_x), sulfur oxides (SO_x), carbon monoxide (CO), carbon dioxide (CO_2), and volatile organic compounds (VOC). Emissions of SO_x are a function of the sulfur content of the fuel used to fire the calciners and the sulfur content of the raw materials used to manufacture ceramics. Emissions of VOC result from incomplete combustion and volatilization of the organic material associated with the raw material. Other beneficiation processes that are associated with emissions include acid leaching and granulation. Emissions of hydrochloric acid (HCl) or other acids may arise from leaching. In addition, PM and products of combustion are emitted from spray dryers used for granulation.

Mixing generally is a wet process. However, VOC emissions from this step may arise from the volatilization of binders, plasticizers, and lubricants. Forming generally is performed in sealed containers and often is a wet process; emissions from this step in the process are likely to be negligible. However, tape casters are a source of VOC emissions. For ceramic bodies that are dryformed, PM is likely to be emitted from grinding, punching, and other green machining activities.

Particulate matter emissions consisting of metal and mineral oxides also arise from glaze preparation, which includes mixing and grinding. Emissions of PM from glaze application also are likely, if the glaze is applied by spraying.

Emissions associated from green ceramic heat treating processes, which include drying, presinter thermal processing, and firing, include combustion products and filterable and condensible PM. Particulate matter emissions consist in part of metals and the inorganic minerals associated with the raw materials. Emissions of the products of combustion are a function of fuel type, raw material constituents, process temperature, and other operating parameters.

Emissions of fluorine compounds also are associated with firing. Fluorine is present in ceramic raw materials in the range of 0.01 to 0.2 percent. As the temperature of green ceramic bodies reaches 500° to 600 °C (930° to 1110 °F), the fluorine in the raw material forms hydrogen fluoride (HF) and other fluorine compounds such as silicon tetrafluoride. Much of the fluorine is released as HF. However, if lime is present in the ceramic body, HF reacts with the lime to form calcium fluoride (CaF_2), thereby reducing potential HF emissions.

Other emission sources associated with ceramics manufacturing include final processing operations and fugitive dust sources. The final processing steps include grinding and polishing, which can emit PM and PM-10, and surface coating, annealing, and chemical treatment, which can emit VOC. Fugitive dust sources, which consist of vehicular traffic, wind erosion of storage piles, and materials handling and transfer, emit PM and PM-10.

Several techniques have been used to control PM emissions from the mechanical processing of ceramic raw materials and finished products. Fabric filters are the most commonly used control device, but wet scrubbers and electrostatic precipitators (ESPs) also are used. Fabric filters, wet scrubbers, and ESPs also are used to control emissions from clay calciners and dryers. Venturi scrubbers and fabric filters are used to control emissions from granulation (spray dryers) and from glaze preparation and application. Afterburners have been used to control VOC emissions from tape casting operations. Emissions from kilns generally are uncontrolled.

Emissions of HF from kilns can be reduced through process modifications such as increasing the raw material lime content and reducing kiln draft, kiln exhaust temperature, and kiln residence time. Dry sorption scrubbing also has been used to control HF emissions in the brick and ceramic industries.

Ceramic Cutting Tools

Ceramic tools have high resistance to heat and to wear and can be used to machine metals that are extremely hard, and they are chemically stable. These attributes allow them to be used to machine metals at high cutting speeds and in dry machining conditions because it is not necessary to reduce temperatures on the cutting edges of these tools.

However, these favorable properties are exchanged in machining for lower toughness when these tools are compared with carbide tools. This deficiency can be offset by selecting an appropriate ceramic cutting grade and the type of tool.

Ceramic tools are based primarily on alumina (Al_2O_3) and silicon nitride (Si_3N_4) compounds and are available in a variety of grades that include ceramics mixed with other materials and reinforcing whisker materials that make them harder.

Oxide Ceramics

Ceramic or “cemented oxide” tools are made primarily from alumina and can contain additions of titanium, magnesium, chromium or zirconium oxides or silicon-carbide grains that are distributed homogeneously throughout the alumina matrix to improve toughness. Cemented oxides are relatively brittle and require highly rigid and vibration-free machines and setups.

The materials are molded at pressure more than 4,000 psi and they are sintered at approximately 3,000 degrees F. This process accounts in part for the high density and hardness of these tools. Oxide ceramics are used primarily for rough and finish turning, for grooving of gray and nodular cast iron and for continuous cutting at high cutting speeds without the use of coolant.

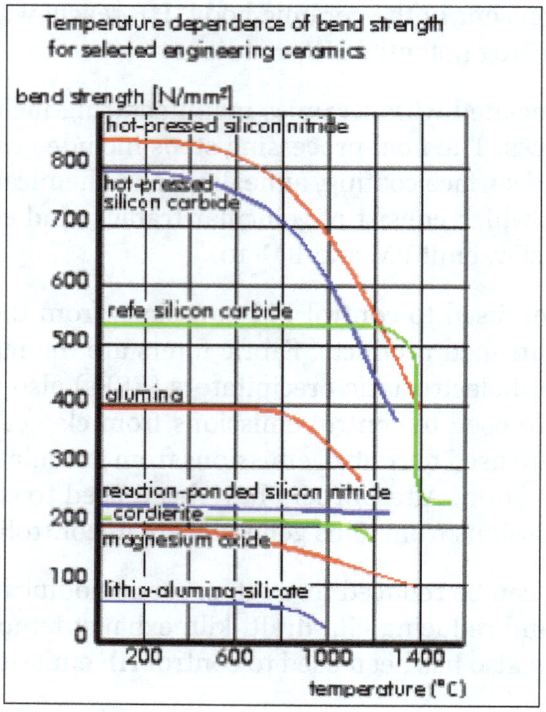

Ceramic tools are stable at extreme temperatures as high as 4,000 degrees F., unlike carbide and cement tools that contain metallic binders and begin to soften at temperatures greater than 1,000 degrees F.

Whiskered Ceramics

Reinforced or whiskered, ceramics use extremely fine-grained silicon-carbide crystals that are called "whiskers" because they resemble small hairs under a microscope to reinforce and toughen basic ceramic compounds.

In ceramic tool materials, single-crystal silicon carbide whiskers, on the order of one micron in diameter and 0.003937 in (100 microns) in length, are intertwined within the alumina-matrix structure. These whiskers have a tensile strength of about 1 million psi and dramatically improve the fracture toughness of the tool material. They also effectively block and prevent propagation of cracks.

Reinforced ceramics work differently from other cutting materials. With reinforced ceramic cutting tools, the objective in machining is to generate high temperatures ahead of the cutting tool to soften or plasticize the workpiece material. That facilitates the removal of material and a reduction in cutting forces. The ideal cutting temperature in nickel alloy is about 1,800 degrees F, for example.

Cutting with ceramic inserts requires high surface speed and balanced feedrates. High speed is necessary to generate the high temperature in the shear zone and to ensure that the heat propagates into the chip-forming zone immediately ahead of the cutter. When cutting speeds are too slow, insufficient heat is generated to soften the material in this zone, and the cutting forces are raised and insert failure occurs.

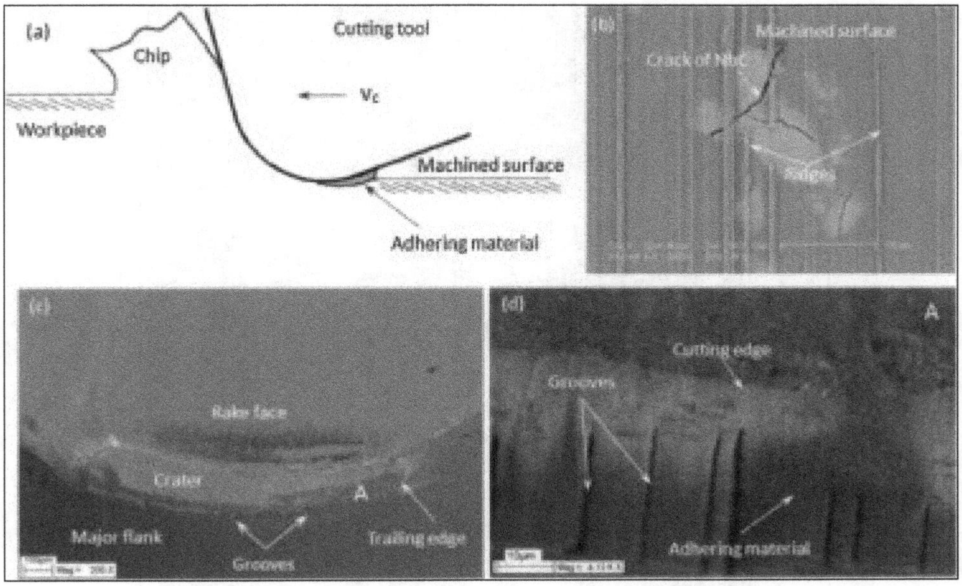

A strategy for using ceramic inserts is to program fewer, but deeper cuts that bury the insert deep in the workpiece. This moves the notch formation further up the face of the insert to an area that has a larger, stronger cross section. Ramping cuts should be programmed to accommodate these tools and fixed depths of cut should be avoided to spread wear over a larger section of the insert.

When machining interrupted cuts with reinforced ceramics, it is important to keep the speed of the cutter high. A rule of thumb is to estimate the percentage of voids in the workpiece surface and increase cutting speed by that percentage. This increase in surface speed offsets the loss of heat generation created by the voids.

Whiskered ceramics work best on hard ferrous materials and difficult-to-machine nickel-base alloys, including Inconel, Waspoloy and Hastelloy. They do not work well on ferrous alloys below Rc 42 hardness because of the chemical reaction that occurs between iron and the carbon that is part of the silicon carbide reinforcing material.

Silicon Nitrides

Ceramics based on silicon nitride offer increased resistance to abrasion and thermal shock and have high fracture toughness. These ceramic tools have a needle-like structure that is embedded in a temperature-resistant grain-boundary. These structures enhance crack deflection, crack bridging and pull-out effects, and lead to superior fracture toughness.

These materials are based on compositions of ceramic powders with high purity. Techniques used to produce these ceramics, including optimized powder processing and gas-sintering, enhance their fracture toughness and high-temperature hardness.

Silicon nitrides are well suited for rough machining cast iron, even under unfavorable conditions such as heavily interrupted cuts and varying depths of cut. Silicon nitrides also are used to mill cast iron, even with positive tool geometries. Silicon nitrides are suitable for milling operations that produce chips with large cross sections and require positive tool geometries.

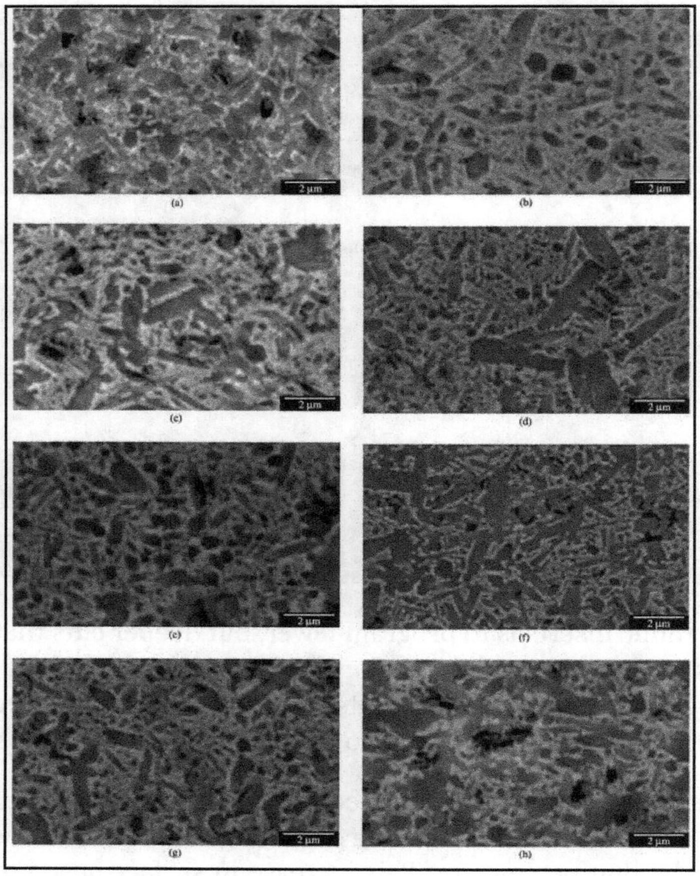

When used with powerful machine tools, silicon nitrides enable high cutting speeds (more than 800 in/min) and feeds (0.2 to 0.3 in/min) for rough boring cast iron. Silicon nitride tools offer fracture resistance, but their relatively low resistance to chemical wear limits their use in machining nodular cast irons. However, wear-resistant, chemical vapor deposit (CVD) alumina coatings have expanded the range of applications for silicon nitride-based tools to include these difficult-to-machine irons. Gray cast iron and nodular graphite cast iron can be milled at cutting speeds of 500 to 800 in/min or sometimes faster than 1,000 in/min.

Five Types of Ceramic Cutting Tools

There are five types os ceramic cutting tools. Here are the basic differences:

Aluminium Oxide or White Ceramic

Pure alumina strengthened by Zirconium:

- High hardness but relatively low strength, toughness and thermal conductivity making them prone to cracking if used with coolant.
- Offer excellent wear characteristic making this range ideal for finish turning, boring and grooving of Cast Iron as well as Tube Scarfing.

Mixed or Black Ceramics

Alumina with metallic phase of Titanium Carbide and Titanium Nitride which improves thermal conductivity.

- Increased ability to cope with extreme cutting conditions.
- Used mainly for hard part machining (up to HRC62).
- Used extensively in Steel and bearing markets and for Cast Iron finishing operations.
- Recent improvements has seen greater wear resistance and toughness due to higher density and reduced porosity.
- The main advantage that mixed ceramic has over Cubic Boron Nitride (CBN) in hard part machining applications is the effect on part production costs. Whilst CBN has a better tool life, the cost per edge of a ceramic insert is a fraction of that of CBN (approximately one twentieth) whilst feed rates (chip thickness around 0.1 mm) and cutting speeds (range between 50-200 m/min) are not significantly higher.

Silicon Nitride based Ceramics

- Low thermal expansion makes this range very resistant to thermal shock.
- High breaking strength sees their use in rough turning (even with heavily interrupted cuts), boring and milling of Cast Iron.
- Offers exceptional tool life.
- Cutting speeds up to 1000 m/min, feeds of 1 mm/rev.
- More stable in operations but tends to suffer from normal flank and crater wear rather than failures associated with white and black ceramics.

Whisker Reinforced Ceramics

Pure alumina structurally strengthens from introduction of Silicon Carbide in the form of whiskers.

- Added tensile strength and fracture toughness even at elevated temperatures.
- Ideal for high speed machining for Nickel based Aerospace alloys, e.g. Inconel 718 correspondingly greater metal rates.

Sialons

Alumina substrate strengthened by Silicon Nitride.

- A lower cost option than Whisker ceramic (but lower performance).
- Widely specified and used in the machining of Aerospace alloys.

Ceramic Coating

Ceramic coating is a clear coat which includes a liquid polymer. When ceramic coating is applied on a car's body, the polymer forms a chemical bonding with the factory paint. Once coated, it can only be removed by abrasion and not by any chemicals. If done correctly, ceramic coating can even last the lifetime of the car.

Ceramic Coating Advantages

Ceramic coating offers good protection to the car's surface. The nano-coating can protect the car from most scratches, dirt and chemical contaminants. Ceramic coating also doesn't have any side-effects to the original paint.

Ceramic coating also lasts longer than regular paint. Sincethe coating is infused with the stock paint due to strong chemical bonding, the protection stays put even under heavy shocks and vibrations.

A car with ceramic coating is also easier to clean. The body panels of a ceramic coated car will be smooth and free of abrasions. Such a surface offers no place for dirt to stick on to. Even if it does, a simple swipe with a clean cloth brings the car back to its radiant finish.

Earlier, cars were given that extra shine by waxing. However, waxing is temporary and will eventually wear o. Both waxing and ceramic coating takes a considerable amount of time to be applied. But in the long run, ceramic coating is better than waxing due to its long life and better finish.

The long life of ceramic coating indirectly makes it highly cost effective. In addition to all the above advantages, ceramic coating makes your car look so much better.

Ceramic Coating Disadvantages

All the above-mentioned advantages of ceramic coating come at a high price. Though it might be cheaper than PPF (Paint Protection Film), the price of ceramic coating is still high.

Ceramic coating application takes time. If you plan to do ceramic coating on your car, be ready to leave it at the detailing shop for at least three days. Also, if not done correctly or professionally, ceramic coating might look worse than the original paint finish of your car.

Ceramic Coating Advantages and Disadvantages

Everybody wants to keep their cars in their best shine all its life. Since ceramic coating offers protection for all the lifetime of the car, people still opt it even at a high cost. There are many reputed detailing shops in India, such as Ceramic Pro, Calgary, and 3M, which offers good quality ceramic coating. If ceramic coating is not your thing, you can always opt for the lesser wax protection or the superior Paint or the superior Paint Protection Film.

Ceramic Machining

Ceramic machining processes involve forming, firing and finishing. Ceramics have a unique combination of mechanical, physical and chemical properties:

- High strength;
- Hardness;
- Low density;
- High stiffness (modulus of elasticity);

- Good tribological properties (e.g., excellent resistance to different types of wear);
- Very low electric conductivity;
- Very low thermal conductivity;
- High refractoriness and thermal stability;
- Good corrosion resistance.

The techniques of manufacturing ceramics (shape forming, sintering) allow to produce the parts of different sizes and shapes with good cost efficiency. However the machining of ceramics is very expensive and time consuming operation representing from 50 to 90% of the total cost of the part.

The main parameter of the efficiency of a machining operation is the Material Removal Rate (MRR) indicating the volume of the material removed from the workpiece surface for 1 min.

The methods of ceramics machining:

- Machining of ceramics in the presintered state;
- Grinding of ceramics;
- Ultrasonic machining of ceramics;
- Rotary ultrasonic machining of ceramics;
- Laser assisted machining of ceramics;
- Laser machining of ceramics.

Machining of Ceramics in the Presintered State

Sintered ceramics are very hard and therefore their machining is an expensive, difficult and time consuming process. Ceramic parts may be effectively machined before the final sintering stage either in the "green" (non-sintered powder) compact state or in the presintered "bisque" state.

Conventional machining methods (milling, drilling, and turning) may be applied for the ceramic parts in the presintered state. Titanium nitride (TiN) coated high speed steel tools, tungsten carbide tools and polycrystalline diamond (PCD) tools are used in machining of presintered ceramics.

The material removal rate (MRR), which may be achieved in machining of ceramics in the bisque (presintered) state is 0.6 in^3/min (9832 mm^3/min). This value is similar or even higher than MRR of Tool and die steels.

Shrinkage and warping of the ceramic part occurring during its sintering does not allow achieving tight dimensional tolerances and high quality surface finish in the presintered state. However machining of bisque ceramics allows reducing the cost of the final machining of the parts in the sintered state.

Grinding of Ceramics

Grinding is the most widely used method of machining of Ceramics in the sintered state. Grinding operation involves a rotating abrasive wheel removing the material from the surface of the workpiece. The grinding zone is continuously flushed with a fluid coolant, which cools the grinding zone, lubricates the contact between the wheel and the part surfaces, removes the micro-chips (debris) produced in the grinding process.

Resin-bond wheels with either synthetic or natural diamond of different grit size pressed at different concentrations in polymer (resin) matrices are commonly used for grinding ceramics. Electrolytic in-process dressing (ELID) technique of dressing metal-bonded grinding wheels is used for fine (nano) finishes grinding. The Material Removal Rate (MRR) of grinding ceramics is maximum 0.0006 in^3/min (9.832 mm^3/min).

Ultrasonic Machining of Ceramics

Ultrasonic machining (UM) of ceramics is the machining method using the action of slurry containing abrasive particles flowing between the workpiece and a tool vibrating at an ultrasonic frequency. The vibration frequency is 19 ~ 25 kHz. The amplitude of vibration 0.0005 - 0.002″ (13 – 50 µm). During the operation the tool is pressed to the workpiece at a constant load.

As the tool vibrates, the abrasive particles (grits) dispersed in the slurry strike the ceramic workpiece and remove small ceramic debris fracturing from the surface. Conventional ultrasonic machining (UM) is characterized by low material removal rates: up to 0.003 in^3/min (49 mm^3/min). Other disadvantages of the conventional ultrasonic machining method are low accuracy and high tool wear. Ultrasonic machining is used commonly for drilling operation.

Rotary Ultrasonic Machining of Ceramics

Rotary ultrasonic machining (RUM) of ceramics combines grinding operation with the method of ultrasonic machining. A core drill tool made of a metal bonded diamond grits is used in the rotary ultrasonic machining (commonly drilling). The tool is rotating and simultaneously vibrating at an ultrasonic frequency. The tool is continuously fed and pressed at a load towards the ceramic workpiece causing abrasive action performed by the rotating-vibrating diamond grits.

A fluid coolant is continuously flowing through the core of the tool to the grinding zone cooling it and removing the debris produced in the grinding process. Rotary ultrasonic machining is much more effective than conventional ultrasonic machining. The RUM material removal rate is up to 0.03 in^3/min (492 mm^3/min).

Laser Assisted Machining of Ceramics

Laser assisted machining (LAM) is the method of machining ceramics using a laser beam directed to the workpiece area located directly in front of the conventional cutting tool. The laser beam heats and softens (not melts) the ceramic material at the surface just prior the cutting action. As a result the cut material becomes ductile and it may be removed much faster than in conventional cutting operation without a laser assistance.

The LAM material removal rate is up to 0.06 in^3/min (983 mm^3/min). Titanium nitride coated tools are used for the laser assisted machining of ceramics. Traditional machining operations (milling, turning) may be performed by the method of the laser assisted machining.

Laser Machining of Ceramics

Laser machining of ceramics is the machining operation performed by a high power laser melting the material, which is blown away by a supersonic gas jet. The laser energy density required for melting alumina ranges from 4.6 BTU/in^2 to 6.1 BTU/in^2 (750 J/cm^2 to 1000 J/cm^2).

The following machining operation may be performed by laser:

- Drilling,
- Cutting,
- Scribing and marking.

Residual stresses and micro-cracks may form at the cut edge as a result of the shrinkage of the solidified molten material. Preheating of the ceramic workpiece to 2550 °F (1399 °C) prior to the laser machining allows to minimize micro-cracking due to reduction of the temperature gradients and thermal stresses.

Ceramic Joining

Ceramics have an important part to play in conjunction with other materials. They can add particular functionality or provide added benefit to a component, e.g. as hard wearing surfaces, ultra-hard materials in cutting tools, for corrosion resistance or high temperature protection. As ceramics are not being used in isolation, joining is an increasingly vital technology for the integration of the materials.

Joining of Ceramics

Despite its obvious importance, joining is often neglected during the design process. Many engineers incorporate ceramics into a component as though they were high performance metals, giving little thought to service conditions or joining operations. This can lead to two outcomes, both the part fails and its designers conclude that the ceramic was unsuitable and that metals should be used as before, or an expensive redesign may be required if a ceramic must be used.

Joining Considerations

There are many important issues to be considered alongside joining such as materials selection, best practice and joint design.

Ceramic Joining Technologies

The ceramic joining technologies used today (few of which have been developed specifically for this class of materials) range from simple mechanical attachment such as the compression fit used in spark plugs, through to liquid phase processes such as adhesive bonding and brazing. The thermal protection system for the space shuttle uses adhesives, and the ceramic turbocharger rotor uses brazing. There are problems associated with these processes including processing considerations (such as component size and joining atmosphere), time constraints and costs, which is why the new ceramic joining technologies are being developed.

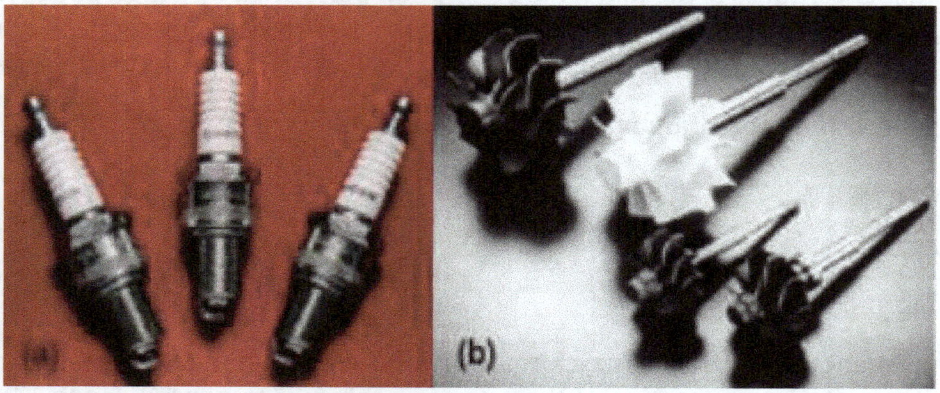

(a) Spark plugs manufactured using compression fitting and (b) turbocharger rotor assemblies made from silicon nitride.

Ultrasonic Joining

Ultrasonic joining, which is used extensively in the plastics industry, has been used for ceramic/metal combinations such as alumina/aluminium, alumina/stainless steel, zirconia/steel and glass ceramic (cordierite-based)/copper. Typical applications include batteries, thread guides, textile cutting equipment and heavy duty electrical fuses. The advantages of the process are the very fast joining times (less than one second), the fact that surface preparation is not critical (contrary to almost every other ceramic joining process) and the lack of melting and intermetallic formation. However, to join hard metals such as steel, soft, deformable interlayers are needed. One limitation of ultrasonic joining of ceramics is that only films or thin sheets of metal can be joined to a ceramic.

Process

Ultrasonic joining requires a transducer assembly operating at about 20 kHz (the source of the ultrasound) coupled to a sonotrode. The sonotrode tip is placed in contact, usually under a clamping load of 1-10 N mm^{-2}, with the workpiece. The heat generated is localised at the interface, creating a temperature of up to 600 °C when using aluminium interlayers.

The bonding mechanism relies on the vibratory shear stress of the metal exceeding its elastic limit, coupled with the breakdown of surface oxide films exposing atomically clean metal. The clamping force exerts plastic deformation on the metal, which increases the interfacial contact between the metal and ceramic. Mechanical keying then occurs across the interface and the joint is formed, perhaps along with some chemical interactions.

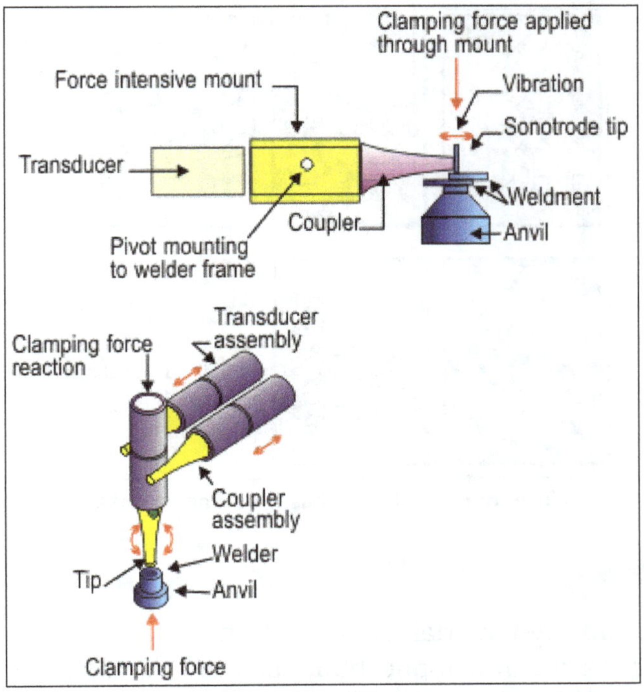

An ultrasonic welding set up.

Applications

In future, ultrasonic joining could be applied to gas-tight container seals (lamps), optical components and joining metallic membranes onto ceramic bodies.

Transient Liquid Phase Bonding

Another technique, transient liquid phase bonding (TLPB), has the ability to produce a bond at a lower temperature than that at which it will be ultimately used. The technology is currently being adapted for a number of ceramics using 'interlayers' based either on glasses (such as oxynitrides for joining SiAlON) or pure metals or alloys (such as Ge and Ge-Si for joining SiC and SiC/SiC composites).

Bonding in the SiAlON system is shown in figure. A mixture of silicon nitride, yttria, silica and alumina are applied by spray coating to one surface of the joint. As the samples are heated to 1600 °C, a load of 2 MNm^{-2} is applied. Joint formation occurs at this temperature over a period of 10-80 minutes. At about 1400 °C, the oxide components react to form a Y-Si-Al-O liquid phase. This leads to densification and sintering. The silicon nitride then dissolves into the liquid, boosting both Si and N contents and altering the composition to Y-Si-Al-O-N. At the same time, (ß-SiAlON grains grow and form an interlocking network across the joint, forming an in-situ reinforcement phase. A secondary process is the diffusion of the adhesive mixture into the surrounding adherent material. Within this diffusion zone the composition and properties of the ceramic gradually change.

The weakness of the TLPB method is that a favorable reaction between the interlayer and the substrates is required. For silicon nitrides, the glassy interlayer must redistribute itself and penetrate the adjoining microstructure.

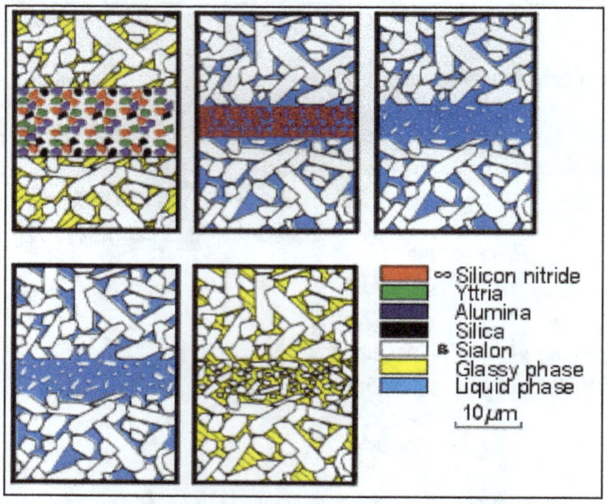

The transient liquid phase bonding process.

Infiltration Processes

High strength, high temperature materials for operation in excess of 1000 °C, such as SiC-based composites, are needed for structural applications such as heat exchangers and gas turbine components. This is because traditional stainless steels and superalloys have reached their operational limits. Industry therefore wants to develop a robust joining process suitable for both SiC monolithic and composite materials. Polymers that react with and infiltrate into the bulk material offer a potential solution.

In these so-called infiltration processes, a mixture of polymer precursor (a source of carbon), aluminium, boron and silicon is applied to the joint surfaces (in a tape, paste or slurry) and then heated, generally to 1200 °C, in an inert atmosphere and air using a propane torch or furnace. The joint forms through pyrolysis of the polymer precursor material, which subsequently reacts with the silicon in the presence of aluminium and boron sintering aids to form in-situ, high density SiC. Strengths range from about 95 MNm^{-2} for air-joined samples down to 40 MNm^{-2} for argon atmosphere samples. The fact that the joint can be produced using a simple gas torch could have a major impact on the repair or on-site production of ceramic/ceramic joints. So far, various grades of SiC, SiC/SiC and C/SiC composites have been joined using this technique.

Microwave Joining

Microwaves provide another technique for joining ceramics together. Microwave energy is already being applied to the drying/firing of refractories and whiteware. Now it is being considered as an energy source for joining ceramics such as alumina, zirconia, mullite, silicon nitride and silicon carbide. The direct coupling of the microwave with the ceramic results in volumetric heating, and so there is great potential for heating large sections uniformly. Control of the location of the maximum electric and magnetic fields also enables precise, selective heating.

Conventional diffusion bonding techniques use radiant heating methods and so the time to reach temperature and the time at bonding temperature can be as long as 8 hours. This is particularly the case for materials such as alumina, which are diffusion bonded at temperatures approaching 1600 °C. Using a microwave heat source, bonding times can be reduced by an order of magnitude.

Very high purity aluminas are difficult to heat, owing to low inherent dielectric properties making joining difficult. Impure, 85% alumina on the other hand is joined easily. The uses of interlayers, including sealing glasses, and alumina gels have both been investigated for producing joints with high purity alumina. Alumina gels offer the advantage that, at the joining temperature, the gel transforms into colloidal α-alumina which subsequently sinters to provide a homogeneous interface. Joints between 85% alumina show bond strengths equivalent to that of the parent material. Joint formation has been studied and a number of possible mechanisms have been identified, depending on the material. For impure aluminas, the glassy grain boundary phase softens and assists in the bonding process, while for zirconia; a solid state process has been identified.

Brazing

Work at TWI is focused mainly on the more traditional methods of joining ceramics, such as brazing, diffusion bonding, glasses and adhesives. However, developments programmes are also investigating the modification of braze alloys by the addition of a ceramic reinforcement. This reinforcement provides a joining medium with a coefficient of thermal expansion (CTE) between that of ceramics and metals, and also gives the joint improved strength owing to the introduction of a second phase i.e. an in situ metal matrix composite. The ability to tailor the CTE of the joining medium is of greater interest. The additional joint strength is a bonus as this raises the possibility of designing the braze to accommodate thermal stresses that would otherwise build up during the joining process.

Sol-gel

For low temperature applications, novel organic based adhesives reinforced with nano-sized ceramic particles are being fabricated at TWI using sol-gel chemistry, a liquid phase process. This technique allows intimate mixing of the ceramic and organic constituents on a molecular scale, producing materials of high purity and with a high level homogeneity. Sol-gel processing provides a joining medium that is tough yet flexible and can be designed to be hydrophobic and self-repairing. Applications for these strong, modified adhesive may be found in the optical, biomedical a defence industries.

Modeling

The difficulty faced in the modeling approach is the lack of reproducible material property data for ceramics and joining media such as brazes, solder adhesives etc. If incorrect data is put in, incorrect joint designs will be produced and will give poor support to the decision making process. A great deal of work is therefore concentrating on the development of appropriate data to plug into the models. Until then, the use of modeling for ceramics joint development will continue to be backed up by a series of practical trials.

Advanced Ceramics Testing

Advanced ceramics consist of metallic oxides, carbides, and nitrides that provide excellent hardness and resistance to high temperatures, wear, and corrosion. They are usually made by mixing powdered minerals with certain chemicals, followed by shaping and heating to form the part. They function as components in electronics, armor, cutting tools, jet engines, medical devices, heat exchangers, and many other products.

Most ceramic materials are metallic oxides, such as aluminum oxide and zirconium oxide. Aluminum oxide is called alumina, its mineral is called corundum, and as a gemstone it is known as sapphire or ruby, second only to diamond in hardness. Synthetic aluminum oxide is equally hard, and is very resistant to abrasion and wear. One of its primary applications is the transparent window for bar code readers in supermarkets.

Zirconium oxide combines strength and toughness with biocompatibility and high resistance to wear. These properties make it suitable for medical implants, particularly in the femoral head of hip implants, and in implants for knees, shoulders, and spines.

Some of the most important ceramics are metallic carbides, such as silicon carbide and tungsten carbide. Silicon carbide applications include heating elements for industrial furnaces, wear-resistant parts for pumps and rocket engines, and semiconductor substrates. Combining silicon carbide with boron carbide and other materials produces lightweight armor that protects people as well as airplanes and helicopters. Tungsten carbide provides superior hardness and strength, making it the material of choice for cutting tools.

Another group of ceramics is composed of nitrides, including silicon nitride and aluminum nitride. Silicon nitride's excellent high temperature strength and fracture toughness make it ideal for aerospace applications such as bearings, bushings, and similar components. Aluminum nitride's unusual combination of excellent thermal conductivity and high electrical resistivity make it uniquely valuable in semiconductor and other electronics applications.

Testing and Ceramics Characterization

Ceramics vary widely in chemistry, function, and shape, and characterization methods are tailored for each component.

Common advanced ceramics currently analyzed at NSL include:

- Alumina (Al_2O_3),
- Boron Carbide (B_4C),
- Boron Nitride (BN),
- Chrome Oxide (Cr_2O_3),
- Magnesium Oxide (MgO),
- Silica (SiO_2),
- Titanium Dioxide (TiO_2),
- Zinc Oxide (ZnO),
- Zirconia (ZrO_2),

- Mixtures (e.g. Silica/alumina or silica/alumina/zirconia).

Examples of analyses include:

- Bulk Analysis (%): SiO_2, Al_2O_3, ZrO_2, CaO, K_2O, Na_2O, MgO, MnO, Fe_2O_3, P_2O_5, ZnO, Cr_2O_3, TiO_2, HfO_2.
- Trace Analysis (ppm): Na, Mg, Al, Si, K, Ca, Ti, Fe, Zn, Se, Zr , Ag, Sn, Sb, Pb, Bi.

Hardness Testing of Ceramics

Hardness is one of the most frequently measured properties of a ceramic. Its value helps to characterize resistance to deformation, densification, and fracture.

Hardness is crucial for cutting tools, wear and abrasion-resistant parts, prosthetic hip-joint balls and sockets, optical lens glasses, ballistic armor, molds and dies, valves, and seals. In fact, many ceramic specifications list minimum hardness requirements. For example, a new ASTM zirconia specification for surgical implants, F 1873-98, stipulates that Vickers hardness (HV) shall be no less than 11.8 GPa (1200 kgf/ mm^2) at a load of 9.8 N (1 kgf).

Although measuring and interpreting ceramic hardness should be routine, pitfalls, controversies, and surprises abound. This article outlines typical hardness measurement techniques, and describes and explains some of the hazards that await the unwary traveler.

Measurement Techniques

Hardness is usually measured on conventional microhardness machines with Knoop or Vickers diamond indenters. These machines make impressions whose diagonal size is measured with an attached optical microscope.

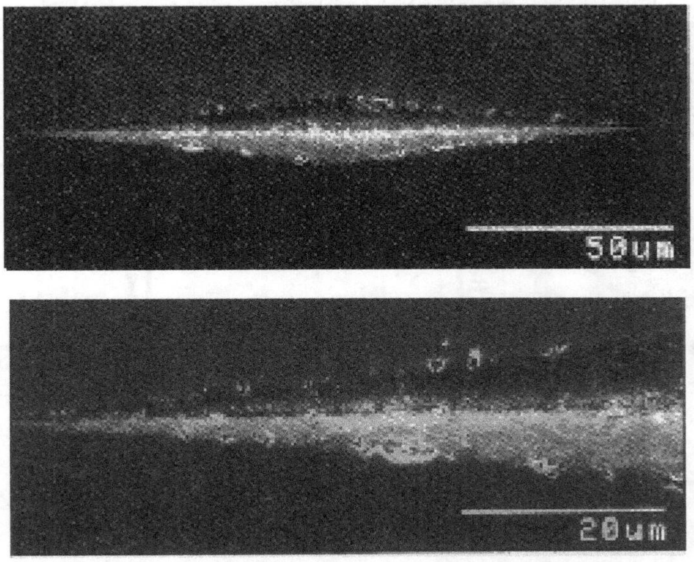

These two images show scanning electron microscope photos of a Knoop indentation (19.6N) in a silicon nitride specimen. Lower photo shows a close-up of the left indentation tip. Slight roughness on the edge of the indentation is caused by individual grain dislodgments.

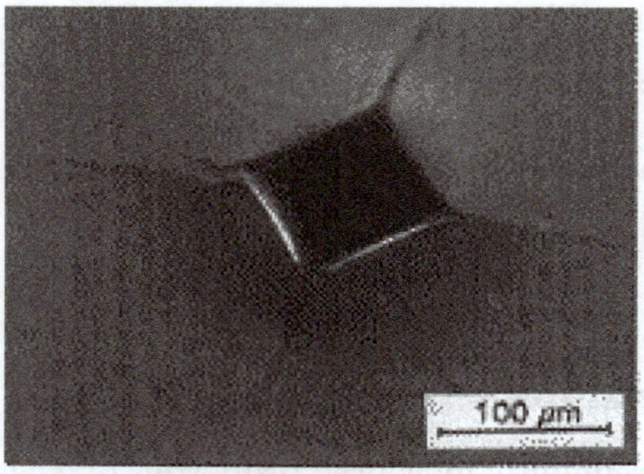

This is an optical photo of a Vickers indentation (98N) in silicon nitride. The specimen is tilted to show the three dimensional form of the indentation, including the tip cracking and the side uplift.

For research purposes, Vickers, Knoop, and Berkovich (triangular pyramid) indenters are customary; Rockwell and Brinell indenters are rarely suitable for ceramics research.

For engineering and characterization applications, approximately 60% of worldwide published ceramic hardness values are Vickers, with loads typically in the range of a few newtons to 9.8 N (1 kgf) and occasional data for soft or high-toughness ceramics as high as 98 N (10 kgf). About 35% are Knoop with loads from as low as 0.98 N (100 gf) to 19.6 N (2 kgf). Knoop hardness is more frequently cited in the United States than in the rest of the world, presumably because of ASTM standards C730 for glass and glass ceramics, and C849 for ceramic whitewares. About 5% are Rockwell, usually either the HRA or superficial HR45N scales.

At low indentation loads, problems arise from the load dependence of hardness and from measurement uncertainty due to the small indentation size. At higher loads, cracking and spalling become problems; in some cases, they make credible measurement impossible. The indentation size effect, in which hardness decreases with increasing indentation load, is found with both Knoop and Vickers hardness. A constant hardness is reached at loads from 5 N to 100 N, depending on the ceramic.

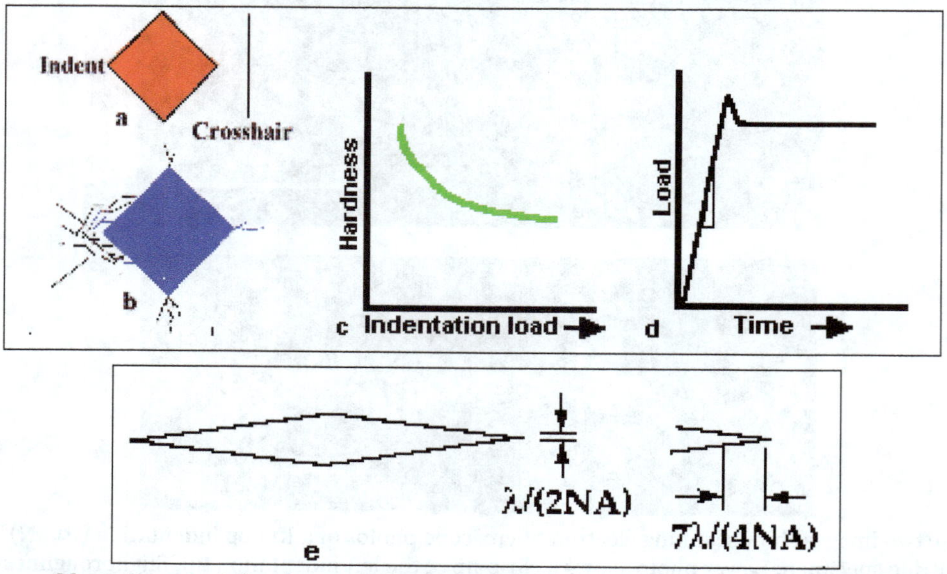

Common problems in ceramic hardness testing include: (a) crosshair technique; (b) indentation cracking; (c) indentation size effect; (d) impact or vibrations; (e) optical resolution limits.

The Knoop hardness often is higher at low loads, but then decreases to a plateau load that is somewhat (-10%) less than the Vickers hardness at high loads. Ideally, the entire hardness versus load curve should be measured, but in practice operators often choose one reference or standard load to allow comparisons between materials. In fact, it is preferable to make the indentations as large as possible to reduce measurement uncertainties, yet not so large as to have the excessive cracking that can interfere with the measurement. The preferred loads for indentation are listed in the following sections. Since hardness is strongly dependent on indentation load, the load should always be reported with the hardness outcome.

Many contemporary structural ceramics have hardnesses in the 10 to 30 GPa range. For the higher hardness, Vickers indentations made at 9.8 N (1 kgf) loads are 25 microns in size and Knoop indentations are 68 microns long. The difficulties in acquiring accurate and precise hardness readings are often not appreciated. Figure shows some of the common, serious issues with hardness measurements in ceramics.

For example, the slightest bump to a hardness machine or the table it sits on while the indenter is in contact with the specimen can create an appreciable error. Cracking compounds the difficulty in estimating where a tip ends, and all too often makes reading the indentation size hopeless.

Furthermore, because hardness is inversely proportional to the square of the diagonal length of the indentation, any error in length measurement is doubled. Therefore, it is crucial that the diagonal length be measured carefully, especially for ceramics in which the indentation size is small and the percentage error may be larger. Remember that smaller diagonal lengths are produced in ceramics than in most metals for a given load, which increases measurement uncertainty. In fact, a recent Versailles Advanced Materials and Standards (VAMAS) round-robin project showed that the reproducibility uncertainty in reported mean hardness of two aluminas was 10 to 15%, and in some instances much higher.

Microscopy Techniques

Correct optical microscopy technique is crucial for ceramic hardness measurements. Reasonable skill, experience, and careful experimental technique are necessary to accurately and precisely measure diagonal lengths. The instrument itself must also meet minimum standards. The early micro hardness literature has many discussions of objective lens quality and design, optical resolution limits, crosshair technique, and especially the need for field and aperture diaphragms to control contrast and brightness. The latter are especially important for translucent or transparent ceramics and glasses. In fact, much of the difference in interpretation of indentation tip location between observers can be traced to illumination and contrast control.

Another aspect of the problem is recognition of the lack of precision in an instrument. Although many machines have digital readouts to 0.1 micron, users should recognize that this is less than the wavelength of light. Therefore, it does not represent the true accuracy or precision, which in fact is probably several times larger. In practice, the resolution of the indentation tips and the subjectivity of the viewer usually lead to a "between-observer" variability of 0.5 microns to 1.0 micron.

Moreover, in spite of the importance of optical techniques, it is amazing that some contemporary commercial hardness machines have no aperture diaphragm, and some have neither an aperture nor a field diaphragm.

Instead, the emphasis with contemporary machines seems to be on capturing the indentation image with a video camera. The image is then projected onto a monitor, and the results are interpreted by a computer. Unfortunately, computer technology cannot overcome the basic optical limitations of an instrument. In many instances, a single monitor pixel can represent as much as 0.5 microns, a significant fraction of the size of a small ceramic indentation. In view of this dimension, it seems unlikely that the tip resolution of such systems can ever match that which is achievable by the trained eye of an expert microscopist.

Knoop Hardness

Frederick Knoop developed his elongated pyramidal indenter as an alternative to the square-base pyramidal Vickers indenter, in large part to overcome the cracking observed in brittle materials. Indeed, experience with a wide range of ceramics has proven that the Knoop indentations are far less likely to crack. Besides ASTM standard E 384, Microhardness of Materials, three material-specific Knoop hardness standards have been developed:

- C 730 for Glass and Glass Ceramics, which recommends a load of 0.98 N (100 gf).
- C 849 for Whitewares, with 9.8 N (1 kgf).
- C 1326 for Advanced Ceramics, with 9.8 N (1 kgf).

The European Community prestandard, CEN ENV 843-3, has both Knoop and Vickers hardness methods as well as Rockwell A and N scales. In addition, a new ISO Ceramic Hardness standard is in preparation in Technical Committee TC 206, Fine Ceramics, and will include both Knoop and Vickers hardness.

Knoop indentations are about 2.8 times longer and are shallower than Vickers indentations made at the same load. In principle, the longer indentations should make an easier-to-read indentation, but in practice the length advantage is offset by the greater difficulty in determining where the tapered tip ends.

A major advantage of the Knoop indentation over Vickers for ceramics is that larger indentation loads may be applied without cracking. With longer indentations, the accuracy and the precision of the length measurements are superior. Even if the sides of the indentation are displaced or cracked, a credible diagonal length reading and hardness estimate may be made.

The tip uncertainty is often of the order of 0.5 mm to 1.0 mm, irrespective of the indentation size; consequently, the percentage error is minimized with long indentations. In addition, it is easier to measure hardness in the constant-hardness region of the indentation size-effect curve. Despite these advantages, Knoop hardness is under-utilized and under-appreciated by the ceramics community.

Optical microscope resolution limits are potentially serious for Knoop indentations due to the slender tapered tip. The error in underestimating the true tip location has been calculated as

$7l/(2NA)$ for Knoop, where l is the wavelength of light and NA is the objective lens numerical aperture. For a typical microscope having a 40X, 0.65 NA objective lens, the calculated correction for green light (l = 0.55 microns) is 3 microns, a significant number.

A correction for this is incorporated in the two older ASTM Knoop standards: C 730 for glass and C 849 for ceramic whitewares, but is not included in the master microhardness of materials standard E 384, or in the advanced ceramic standard, C 1326. The confusion about whether to add this correction factor has reached the point where a major glass manufacturer has listed both uncorrected and corrected numbers for Knoop hardness numbers in its product data handbooks.

To help determine whether the full $7l/(2NA)$ correction is appropriate, NIST compared calibrated scanning electron microscope diagonal length measurements to optical microscope measurements on a single SEM-measured Knoop impression in a silicon nitride reference block. The indentation was measured optically by four skilled operators. Three used the optical system on a conventional microhardness machine, and the fourth used a metallograph that is applied in certifying metallic microhardness SRMs.

The length was measured through several lenses with different numerical apertures, all calibrated by the same NIST-certified stage micrometer. Five or ten repetitions were made by each observer. Tne SEM measurements benefited from higher magnification photographs (1500X, 5000X) to aid in tip location. The mean SEM length measurement was 146.8 microns (+0.2 microns). The mean optical diagonal lengths were 0.4 to 2.1 microns shorter than the SEM readings. These differences are less than the full $7l/(2NA) = 3.0$ micron correction, probably because the optical observers discerned the tip as a faint black line, albeit not necessarily as two distinguishable tip lines.

Later, as part of an eleven-laboratory international round robin, three certification labs obtained average diagonal length readings (10 indentations) that were within 0.4 to 1.2 microns of the NIST SEM readings. This is better than 1% agreement on 142 micron long indentations, and it underscores how the percentage error may be kept small by utilizing large Knoop indentation sizes.

Vickers Hardness

The square pyramidal indenter creates smaller, deeper impressions that are more likely to crack than Knoop indentations. ASTM standard E 384, Microhardness of Materials, covers Vickers hardness; C 1327 is a new standard for Vickers hardness of advanced ceramics and recommends a load of 9.8 N (1 kgf). Vickers indentations are rarely made for hardness measurements in glasses, and no ASTM or ISO standards have been written for glass. The optical resolution limits are estimated to be only $1.0l/2(NA)$ or ~0.4 microns for Vickers indentations.

Hardness SRM

The poor results from a 1988-1989, VAMAS round robin study with Knoop, Vickers, and Rockwell tests on alumina ceramics underscored the need for standard reference materials (SRMs). In 1994, NIST commenced a project for SRM 2830, Knoop Mardness of Ceramics, which is now available from NIST. It is a HIP'ed silicon nitride block that was prepared from a ceramic bearing ball. It has a high-quality polish with five well-defined indentations, and has a nominal hardness of 13.7 GPa (1400 kgf/ mm^2). A typical 19.6 N impression is shown in figure.

The average diagonal length (~142.0 microns) for each block is listed and certified to within 0.6 microns (0.4%) at a 95% confidence interval. The 19.6 N loads is needed to exploit the advantage that long Knoop indentations provide in reducing the percentage error to such remarkably low levels. Hardness is certified to within 0.9% or within 0.12 GPa (12 kgf/mm²). A calibrated scanning electron microscope was chosen to make all length measurements. A subsequent eleven-laboratory, international round robin verified that operators with conventional optical microscopes produced readings in excellent agreement with the SEM readings. This was especially true for the three participating certification labs: NIST's Metallurgy Division, Wilson Division of Instron, and the Materials Testing Institute, Nord Rhein-Westfalen, Germany.

SRM 2831, Vickers hardness, is still in preparation. It will be a tungsten carbide block and will have five indentations made at a load of 9.8 N (1 kgf). It was difficult to find an opaque ceramic that did not crack at the tips in a fashion that would interfere with the length measurement.

Fracture Toughness

Testing operators usually seek to avoid cracking that interferes with the hardness measurement. However, the ceramic community has contrived a simple method to estimate fracture toughness (K_{Ic}) from the length of the cracks that do emanate from the corners of a Vickers indentation. The lengths of the cracks and the indentation half-diagonal size are related to the hardness, elastic modulus, and fracture toughness by an analytical expression. The early work on this methodology claimed that calculated K_{Ic}'s were accurate to within 30 to 40%, a considerable uncertainty. The mediocre fit success of the early equations prompted the ceramics community to spawn a plethora of several dozen alternative expressions, to the point that massive confusion now reigns.

Unfortunately, the method is limited because toughness depends on the measured crack length raised to the 1.5 power. The substantial errors and uncertainties in measuring the crack size (far worse than measuring indentation size) are thus magnified. In fact, another VAMAS round robin demonstrated variability of a factor of almost two in reported toughness. The requirement to develop crack lengths that are sufficiently long (>2.0X the half diagonal size) has led some to apply enormous loads, sometimes up to 500 N. This often causes severe shattering, prompting one skeptical observer to remark that indentations in some materials might resemble "nuclear bomb craters."

Further confusion ensued when the indentation-fracture community confounded the world by sometimes calculating a Vickers hardness based on load over pojected surface area or $H = 2.0\ P/d^2$. This is in contrast to the worldwide convention in which HV is defined as load over contact area or $HV = 1.8544\ P/d^2$, an expression used in every single Vickers hardness standard on this planet.

Lamentably, years of work and tens of thousands of experiments have still not refined the ~30% uncertainty estimate for indentation fracture toughness. Thus, this method is unsatisfactory for accurate results.

Instrumented Hardness

Instrumented hardness testing, in which displacement and load are recorded simultaneously during load application or removal, is an important emerging technology. Both Vickers and Berkovich (triangular pyramid) indenters are suitable. Various indices of hardness may be deduced from the load and depth of penetration of the indenter. An enormous advantage of this methodology is that it obviates the need for microscopy for measuring the indentation size, thereby eliminating operator skill or subjectivity, as well as microscopy limitations.

On the other hand, complications arise on account of the need to make assumptions about the analytical form of the load-displacement curves. Unloading curves may be analyzed to determine the elastic modulus, but again, assumptions about the indenter shape and penetration geometries must be made. No consensus has been formed relating to the interpretation of these curves.

As a result, for example, one commercial apparatus that is widely preferred in Europe for measuring so-called "universal hardness," actually measures a hardness defined as the load divided by assumed contact area while the load is still applied. Consequently, this universal hardness includes both plastic and elastic deformation components.

Sometimes, ordinary microhardness machines may be retrofitted with load and displacement transducers, but care must be taken to measure displacement as close to the indenter-specimen contact point as possible, and machine compliance is an important interference. Severe variability (up to a factor of four) in instrumented hardness results were recently demonstrated in a VAMAS round robin exercise with a borosilicate crown glass and NIST's SRM 2830 silicon nitride blocks. Dedicated low load instrumented hardness or even extremely low-load nanohardness machines are now commercially available. This emerging technology has great promise, but consensus on the analyses, standard procedures, and reference materials is sorely needed.

Size effect and Brittleness

Close examination of ceramic Vickers hardness/load curves suggests that a discrete transition point may be found on either HV versus load or HV versus diagonal size graphs. At this point, hardness changes from being load-dependent to a constant value, as illustrated in figure.

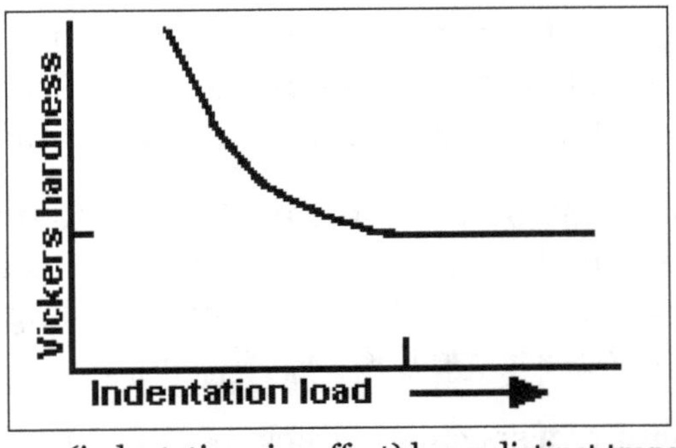

The hardness versus load curve (indentation size effect) has a distinct transition to a plateau hardness.

The transition point appears to be associated with the onset of extensive cracking around and underneath the indentation. Cracking, an integral response of the ceramic to indentation even at small loads, may either be localized or, at higher loads, massive to the extent that crushing follows. A recent paper analyzes the transition point and relates it to a new index of ceramic brittleness, which is defined as $B=HE/(K_{1c})^2$. The brittleness index may be important for predicting wear or erosion resistance, or for the characterization of machinability.

Fracture Toughness Tests of Ceramics

Fracture is a process of breaking a solid into pieces as a result of stress. There are two principal stages of the fracture process:

- Crack formation,
- Crack propagation.

There are two fracture mechanisms: ductile fracture and brittle fracture. Ceramic materials have extremely low ductility, therefore they failure by brittle mechanism.

- Brittle fracture,
- Fracture toughness,
- Flexure test,
- Indentation fracture test.

Brittle Fracture

Brittle fracture is characterized by very low Plastic deformation and low energy absorption prior to breaking. A crack, formed as a result of the brittle fracture, propagates fast and without increase of the stress applied to the material. The brittle crack is perpendicular to the stress direction.

There are two possible mechanisms of the brittle fracture: transcrystalline (transgranular, cleavage) or intercrystalline (intergranular). Cleavage cracks pass along crystallographic planes through the grains. Intercrystalline fracture occurs through the grain boundaries, embrittled by segregated impurities, second phase inclusions and other defects. The brittle fractures usually possess bright granular appearance.

Fracture Toughness

Fracture Toughness is ability of material to resist fracture when a crack is present. The general factors, affecting the fracture toughness of a material are: temperature, strain rate, presence of structure defects and presence of stress concentration (notch) on the specimen surface.

Stress-intensity Factor (K) is a quantitative parameter of fracture toughness determining a maximum value of stress which may be applied to a specimen containing a crack (notch) of a certain length. Depending on the direction of the specimen loading and the specimen thickness, four types of stress-intensity factors are used: K_C, K_{IC}, K_{IIC}, K_{IIIC}.

- K_C: Stress-intensity factor of a specimen, thickness of which is less than a critical value. K_C depends on the specimen thickness. This condition is called plane stress.
- K_{IC},K_{IIC}, K_{IIIC}: stress-intensity factors, relating to the specimens, thickness of which is above the critical value therefore the values of K_{IC}, K_{IIC}, K_{IIIC} do not depend on the specimen thickness. This condition is called plane strain.
- K_{IIC} and K_{IIIC}: stress-intensity factors relating to the fracture modes in which the loading

direction is parallel to the crack plane. These factors are rarely used for metallic materials and are not used for ceramics.

- K_{IC}: plane strain stress-intensity factor relating to the fracture modes in which the loading direction is normal to the crack plane. This factor is widely used for both metallic and ceramic materials.

K_{IC} is used for estimation critical stress applied to a specimen with a given crack length:

$$\sigma_C \le K_{IC}/(Y(\pi a)^{1/2})$$

where,

- K_{IC} – stress-intensity factor, measured in MPa*m½;
- σ_C– the critical stress applied to the specimen;
- a – the crack length for edge crack or half crack length for internal crack;
- Y – geometry factor.

Two test methods are used for measuring fracture toughness parameter (stress-intensity factor) of ceramic materials: Flexure Test and Indentation Fracture Test.

Flexure Test

The test method is similar to that which is used for measuring Flexural Strength, however notched specimens are used.

Indentation Fracture Test

Vickers Hardness Method is used for this test. Polished surface of a ceramic sample is indented by Vickers Indenter, resulting in formation of four cracks emanating from the indent corners. The cracks length is inversely proportion to the material toughness; therefore K_{IC} may be estimated by measuring the cracks length.

References

- Five-types-of-cermamic-cutting-tools, technical: nextgentooling.com, Retrieved 9 May, 2019
- Ceramic-coating-advantages-and-disadvantages, four-wheelers: drivespark.com, Retrieved 31 March, 2019
- Machining-of-ceramics: substech.com, Retrieved 14 July, 2019
- Ceramics, materials: nslanalytical.com, Retrieved 17 May, 2019
- Ceramics: metallography.com, Retrieved 19 April, 2019
- Fracture-toughness-tests-of-ceramics: substech.com, Retrieved 5 February, 2019

Permissions

All chapters in this book are published with permission under the Creative Commons Attribution Share Alike License or equivalent. Every chapter published in this book has been scrutinized by our experts. Their significance has been extensively debated. The topics covered herein carry significant information for a comprehensive understanding. They may even be implemented as practical applications or may be referred to as a beginning point for further studies.

We would like to thank the editorial team for lending their expertise to make the book truly unique. They have played a crucial role in the development of this book. Without their invaluable contributions this book wouldn't have been possible. They have made vital efforts to compile up to date information on the varied aspects of this subject to make this book a valuable addition to the collection of many professionals and students.

This book was conceptualized with the vision of imparting up-to-date and integrated information in this field. To ensure the same, a matchless editorial board was set up. Every individual on the board went through rigorous rounds of assessment to prove their worth. After which they invested a large part of their time researching and compiling the most relevant data for our readers.

The editorial board has been involved in producing this book since its inception. They have spent rigorous hours researching and exploring the diverse topics which have resulted in the successful publishing of this book. They have passed on their knowledge of decades through this book. To expedite this challenging task, the publisher supported the team at every step. A small team of assistant editors was also appointed to further simplify the editing procedure and attain best results for the readers.

Apart from the editorial board, the designing team has also invested a significant amount of their time in understanding the subject and creating the most relevant covers. They scrutinized every image to scout for the most suitable representation of the subject and create an appropriate cover for the book.

The publishing team has been an ardent support to the editorial, designing and production team. Their endless efforts to recruit the best for this project, has resulted in the accomplishment of this book. They are a veteran in the field of academics and their pool of knowledge is as vast as their experience in printing. Their expertise and guidance has proved useful at every step. Their uncompromising quality standards have made this book an exceptional effort. Their encouragement from time to time has been an inspiration for everyone.

The publisher and the editorial board hope that this book will prove to be a valuable piece of knowledge for students, practitioners and scholars across the globe.

Index

S

T

V

W

Y

Z

www.ingramcontent.com/pod-product-compliance
Lightning Source LLC
LaVergne TN
LVHW080647210926
848485LV00003B/343
* 9 7 8 1 6 3 2 3 8 8 6 7 4 *